Reveal MATH®

Student Edition
Grade 4 • Volume 1

McGraw Hill

Back cover: Ray Wise/Moment/Getty Images

mheducation.com/prek-12

Contents in Brief

Welcome to *Reveal Math!*

We are excited that you have made us part of your math journey.

Throughout this school year, you will explore new concepts and develop new skills. You will expand your math thinking and problem-solving skills. You will be encouraged to persevere as you solve problems, working both on your own and with your classmates.

With *Reveal Math*, you will experience activities to spark curiosity and challenge your thinking. In each lesson, you will engage in sense-making activities that will make you a better problem solver. You will have different learning experiences to help you build understanding.

We look forward to revealing to you the wonder and excitement of math.

The *Reveal Math* authors

The *Reveal Math* Authorship Team

McGraw-Hill teamed up with expert mathematicians to create a program centered around you, the student, to make sure each and every one of you can find joy and understanding in the math classroom.

Ralph Connelly, Ph.D.
Authority on the development of early mathematical understanding.

Annie Fetter
Advocate for students' ideas and student thinking that fosters strong problem solvers.

Linda Gojak, M.Ed.
Expert in both theory and practice of strong mathematics instruction.

Sharon Griffin, Ph.D.
Champion for number sense and the achievement of all students.

Ruth Harbin Miles, Ed.S.
Leader in developing teachers' math content and strategy knowledge.

Susie Katt, M.Ed.
Advocate for the unique needs of our youngest mathematicians.

Nicki Newton, Ed.D.
Expert in bringing student-focused strategies and workshops into the classroom.

John SanGiovanni, M.Ed.
Leader in understanding the mathematics needs of students and teachers.

Raj Shah, Ph.D.
Expert in both theory and practice of strong mathematics instruction.

Jeff Shih, Ph.D.
Advocate for the importance of student knowledge.

Cheryl Tobey, M.Ed.
Facilitator of strategies that drive informed instructional decisions.

Dinah Zike, M.Ed.
Creator of learning tools that make connections through visual-kinesthetic techniques.

Math Is...

Generalize Place-Value Structure

Addition and Subtraction Strategies and Algorithms

Multiplication as Comparison

Numbers and Number Patterns

Multiplication Strategies with Multi-Digit Numbers

Unit 7

Division Strategies with Multi-Digit Dividends and 1-Digit Divisors

Let's Talk About Math!

Throughout this year, you will explore the language of mathematics as you talk about math with your classmates. You are going to learn many new words this year. Use these resources as you expand your vocabulary.

Glossary

In the back of this book, you will find a glossary with definitions for your reference.

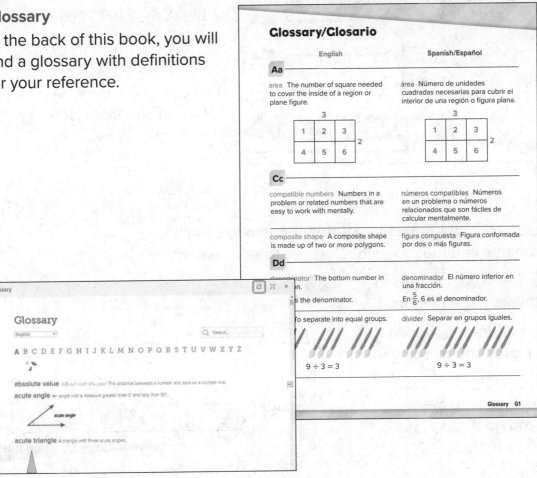

Interactive Glossary

The Interactive Glossary will support you as you work through your Interactive Student Edition and complete assignments online.

Jump into Learning!

You can find all the resources you need from your **Student Dashboard**.

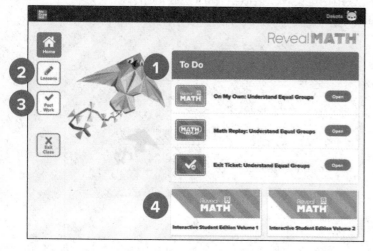

1. Easily access scheduled work or assessments from the To-Do List.

2. View specific lesson resources throughout the course.

3. Review the previously completed work and see your scores.

4. Access the Interactive Student Edition and the eToolkit easily with quick links.

You can use your **Interactive Student Edition** to complete assignments and practice and reference lesson content.

1. Use the slide numbers to find your page number.

2. Type or draw to work out problems and respond to questions.

3. Check your answers as you go through your assignment.

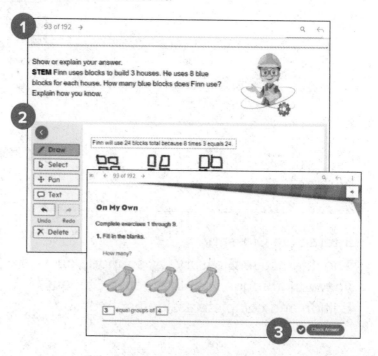

Access Lesson Supports Online!

In addition to your Interactive Student Edition, access these supports online while you practice.

Need an Instant Replay of the Lesson Content?

Math Replay videos offer a 1–2 minute overview of the lesson concept to use as a reference while you are practicing or completing your homework.

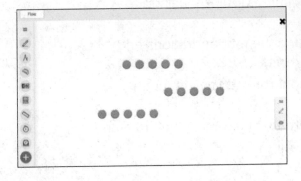

Virtual Tools to Help You Problem Solve

You can access the eToolkit at any time from your Student Dashboard. You will have access to the following manipulatives:

- Counters
- Base-Ten Blocks
- Array Builder
- Fraction Model
- Bucket Balance

- Geometry Sketch
- Money
- Fact Triangles
- Number Line
- and more!

Key Concepts and Learning Objectives

Key Concept **Habits of Mind and Classroom Norms for Productive Math Learning**

- I make sense of problems and think about numbers and quantities. (Unit 1)

- I share my thinking with my classmates. (Unit 1)

- I can use math to make sense of everyday problems. (Unit 1)

- I see patterns in math. (Unit 1)

- I describe my math story. (Unit 1)

- I work productively with my classmates. (Unit 1)

Key Concept **Place Value, Multi-Digit Arithmetic, and Properties of Operations**

- I can read and write numbers up to one million in multiple forms. (Unit 2)

- I can round multi-digit numbers to any place-value position. (Unit 2)

- I can add and subtract whole numbers within 1,000,000 using the standard algorithm. (Unit 3)

- I can solve multi-step word problems using the four operations and assess the reasonableness of answers. (Units 3, 4, 6, 7)

- I can distinguish multiplicative comparison from additive comparison. (Unit 4)

- I can generate a number or shape pattern that follows a given rule and identify apparent features of the pattern that were not explicit in the rule itself. (Unit 5)

- I can find all factor pairs for a whole number from 1–100. (Unit 5)

- I can determine whether a given whole number in the range 1–100 is a multiple of a given one-digit number. (Unit 5)

- I can determine whether a given whole number in the range 1–100 is prime or composite. (Unit 5)

- I can multiply a whole number of up to four digits by a 1-digit whole number and multiply two 2-digit numbers. (Unit 6)

- I can find whole-number quotients and remainders with up to four-digit dividends and one-digit divisors. (Unit 7)

Key Concept Fractions

- I can use fraction models to explain why two fractions are equivalent and generate equivalent fractions. (Unit 8)
- I can compare two fractions using benchmark fractions or by generating equivalent fractions. (Unit 8)
- I can decompose a fraction or mixed number into a sum of fractions with the same denominator. (Units 9, 10)
- I can add and subtract fractions and mixed numbers with like denominators. (Units 9, 10)
- I can multiply a fraction or a mixed number. (Unit 11)
- I can represent fractions with denominators of 10 or 100 using decimal notation and compare two decimals to hundredths. (Unit 12)
- I can add fractions with denominators 10 and 100 by using equivalent fractions. (Unit 12)

Key Concept Measurement and Data

- I can convert larger units of measurement to smaller equivalent units. (Unit 13)
- I can determine and apply the formulas for the area and perimeter of a rectangle. (Unit 13)
- I can display and interpret measurement data in line plots to solve problems. (Unit 13)

Key Concept Analyze and Classify Geometric Shapes

- I can identify and draw points, lines, line segments, and rays. (Unit 14)
- I can classify angles as right, acute, or obtuse, and measure and draw angles. (Unit 14)
- I can draw perpendicular and parallel lines and identify them in 2-dimensional figures. (Unit 14)
- I can recognize that when an angle is decomposed into parts, the angle measure of the whole is the sum of the angle measure of the parts. (Unit 14)
- I can classify 2-dimensional figures by the presence or absence of parallel and perpendicular lines, or the presence or absence of angles of a specified size. (Unit 14)
- I can recognize a line of symmetry for a 2-dimensional figure. (Unit 14)
- I can explain how to find lines of symmetry on 2-dimensional figures. (Unit 14)

Math is...

How would you complete this sentence?

Math is.....

Math is not just carrying out operations and solving equations.

Math is...

- working together
- finding patterns
- sharing ideas
- listening thoughtfully to our classmates
- sticking with a task even when it is a little challenging

In *Reveal Math*, you will develop the habits of mind that strong doers of math have. You will see that math is all around us.

Let's be Doers of Mathematics

Remember, math is more than getting the right answer. It is a tool for understanding the world around you. It is a language to communicate and collaborate. Be mindful of these prompts throughout the year to access the power of math.

1. **Math is... Mine**
   - Mindset

2. **Math is... Exploring and Thinking**
   - Planning
   - Connections
   - Thinking

3. **Math is... in My World**
   - In My World
   - Modeling
   - Choosing Tools

4. **Math is... Explaining and Sharing**
   - Explaining
   - Sharing
   - Precision

5. **Math is... Finding Patterns**
   - Patterns
   - Generalizations

6. **Math is... Ours**
   - Mindset

Lesson **3-1**
Understand Equal Groups

?

Be Curious

What do you notice?
What do you wonder?

Math is... Mindset
What can you do to be an active listener?

Math is... Mindset

What can you do to be an active listener?

Explore the Exciting World of STEM!

Ever wonder how math applies in the real world? In each unit, you will learn about a STEM career that engages in mathematics to make a positive impact in society, from protecting our parks to exploring outer space. Throughout the unit, you will have opportunities to dig into the STEM career through digital simulations and projects.

STEM Career Kid: Meet Hiro
Let the STEM Career Kid introduce their career and talk about their respective job responsibilities.

cinoby/Creatas Video/Getty Images

Math In Action: Ocean Engineer
Watch the Math in Action to see how the math you are learning applies to the real world and help problem solve.

Hi, I'm Hiro.

I want to know everything about our oceans. The ocean has amazing plants and animals. I want to be an ocean engineer when I grow up to make sure our oceans are protected and everyone can enjoy them.

Math Is ...

Focus Question

What does it mean to do math?

Hi, I'm Dakota.

This is going to be a great year! We will learn a lot of math and see how math helps us understand our world. Look out the window. Where do you see math?

Math Tools

Crafts

STEM video | GO ONLINE

Name _____

Let's Shake

1. Five strangers meet. Each person shakes hands with everyone else in the group. How many handshakes will there be?

2. Six strangers meet. Each person shakes hands with everyone else in the group. How many handshakes will there be?

3. Ten strangers meet. Each person shakes hands with everyone else in the group. How many handshakes will there be?

Be Curious

What do you notice?
What do you wonder?

Learn

Math gives us power to solve problems and do all sorts of other good things. Every person has skills in math. Math skills help us grow and reach our goals.

Your teacher has special math skills.

What are your special math skills or math superpowers?

Math is... Mindset

What makes me special in math?

How do your math skills help you?

Math is... Mindset

How can I use my math skills?

What math skills or superpowers do you have that others might not know about?

Math is... Mindset

What are my strengths in math?

What new math skills do you want to develop?

Math is... Mindset

What do I want to learn about math?

Work Together

Think of some other math skills or superpowers that someone might have.

Name

What is your math superpower?

 **Reflect**

What about my "Math Me" do I want someone else to know?

Be Curious

What do you notice?
What do you wonder?

Learn

Aziza cut the brownies she made into 4 rectangular pieces.

What could be the dimensions of the pieces?

12 in.

8 in.

 When we do math, we use many strategies to make sense of problems.

I know

- Aziza baked brownies in an 8 in. by 12 in. pan.

- She cut the brownies into four pieces.

- The four pieces are rectangular in shape.

Math is... Exploring

What do I know about the problem?

I can ask

- How might I represent the problem?

- How can I show how Aziza cut the brownies?

- Did she cut the brownies into pieces of equal size?

- How can the dimensions of the pan help me answer the question?

Math is... Planning

What questions can I ask myself?

When we do math, we work to solve problems, but sometimes the first try doesn't work. If the first try doesn't work, we keep trying and don't give up.

- I can use a different representation.

- I can try to find an equation.

- I can try using different numbers.

Math is... Perseverance

What can I do if I'm stuck?

When we do math, we think about numbers in all sorts of ways.

How can I think about each piece of the brownie?

- If each piece is the same part of the whole, the dimensions could be 6 inches by 4 inches.

- If each piece is a different part of the whole, the pieces could be different sizes.

When we do math, we think about how numbers and quantities relate.

Can the numbers represent the same quantity?

- If I cut into four equal pieces, the numbers will represent the same part of the whole.

- Each is $\frac{1}{4}$ of the whole.

Work Together

Aziza's mother baked brownies in a pan that is 6 inches by 12 inches. She cut the brownies into 4 pieces. What could be the dimensions of the pieces if she cut into equal-sized pieces? What fraction of the whole is each piece? What could be the dimensions of the pieces if she did not cut into equal-sized pieces?

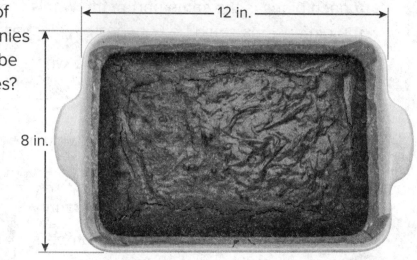
Aziza made another batch of brownies. She cut the brownies into six pieces. What could be the dimensions of the pieces?

12 in.

8 in.

Reflect

Tell about a time when you had a problem and you didn't give up. It might be a math problem. But it might be a problem you had at home, playing a game, playing a sport, playing an instrument, drawing a picture, or doing a puzzle.

Be Curious

What do you notice?
What do you wonder?

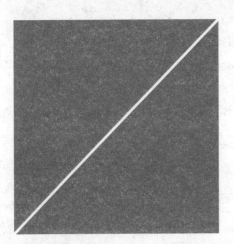

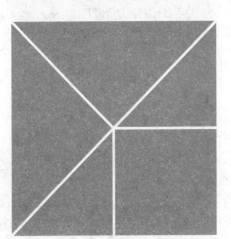

Learn

A baker cuts his breads into pieces of different sizes every day. His Monday bread is shown. Jung bought piece 1 and Han bought piece 2.

How much of the whole is Jung's piece?
How much of the whole is Han's piece?

Whole Grain Breads
$16

▰ When we do math, we make models to represent the problem situation. We then draw conclusions from models to solve problems.

I can use drawings as a model to visualize the problem. The bread is cut into six pieces.

- Two pieces are the same size.

- Four pieces are the same size.

Math is... **In My World**

How can I visualize the problem?

I can use drawings to represent the problem.

- I can fit two more long pieces on the whole. Each long piece is $\frac{1}{4}$ of the loaf of bread.

Math is... **In My World**

How can I represent the problem?

When we do math, we use different tools.

I can use fraction strips to represent and solve the problem.

- Four long pieces make a whole loaf of bread.
- Each long piece is $\frac{1}{4}$ of the loaf.

- Eight short pieces make a whole loaf.
- Each short piece is $\frac{1}{8}$ of the loaf.

1							
$\frac{1}{8}$	$\frac{1}{8}$	$\frac{1}{8}$	$\frac{1}{8}$	$\frac{1}{8}$	$\frac{1}{8}$	$\frac{1}{8}$	$\frac{1}{8}$

The whole loaf is cut into pieces, so fraction strips are a good tool to represent and solve the problem.

Math is... Choosing Tools

What tool can I use to represent the problem?

Math is... Choosing Tools

How do I know this is the best tool to solve the problem?

Work Together

On Wednesday, the baker cuts the loafs of bread into the pieces shown. What part of the whole are pieces 1, 2, and 3?

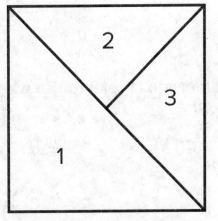

On My Own

Name

Friday's bread is shown. What part of the whole is each piece?

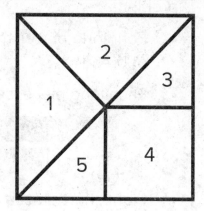

Reflect

How can I represent a problem in math?

What tools do I prefer to use when solving problems involving fractions?

Be Curious

How are they the same?
How are they different?

Learn

Ms. Jenoba has $45.
She will give the same amount to each of her grandchildren.
Each grandchild will get a whole dollar amount.

How many grandchildren might she have?
How much will each grandchild receive?

◤ When we do math, we have to defend our thinking. Sometimes we use words. Sometimes we use numbers and pictures.

I can use words, equations, or drawings to explain my thinking.

Ms. Jenoba might have:

- $9 to give to each grandchild, so she has 5 grandchildren.

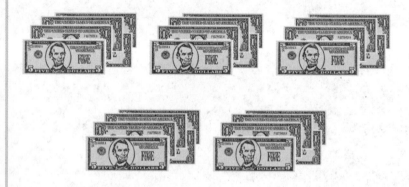

Math is... **Exploring**

How can I explain
my thinking?

◤ When we do math, we listen to the arguments of others and think about what makes sense and what doesn't.

Someone else might think that Ms. Jenoba has:

- 3 grandchildren, so she gives each $15.

Math is... **Sharing**

What can I learn from
others' thinking about
the problem?

◤ When we do math, we know when an estimate or an exact answer is appropriate.

I need an exact answer.

- Ms. Jenoba has $45 to share equally.

- All grandchildren get whole dollar amounts.

Math is... Precision

Do I need an exact answer or an estimate?

◤ When we make arguments, we try to be precise.

We use correct vocabulary and make sure our calculations are accurate. We label our drawings and include units of measurement.

I used exact numbers in my argument.

I also used appropriate labels.

- Ms. Jenoba gave $9 to each of 5 grandchildren.

I checked my calculations and they are accurate.

- $5 \times 9 = 45$

Math is... Precision

Is my argument clear and exact?

💬 Work Together

Ms. Jenoba now has $48 for her grandchildren. Each will receive the same whole dollar amount. How many grandchildren might she have? How much will each receive?

On My Own

Name _____

Mr. Major has $36.

He will give the same amount to each of his children.

Each child will get a whole dollar amount.

How many children might he have?
How much will each child receive?

Reflect

Why is it important to listen to and hear everyone's ideas?

Be Curious

**What do you notice?
What do you wonder?**

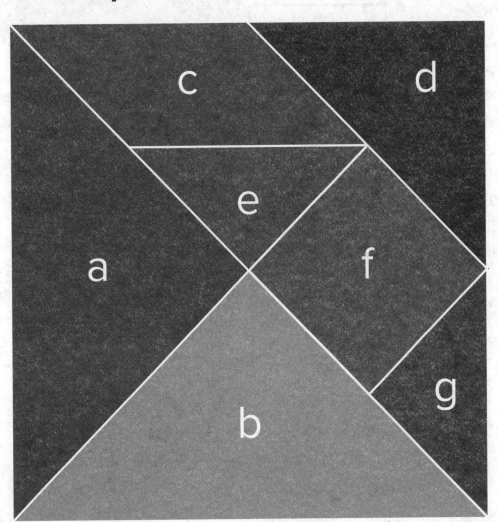

Learn

The large square represents one whole.

What fraction of the whole would each of the smaller pieces represent?

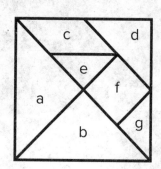

◀ Math is full of patterns and relationships. When we do math, we look for patterns and relationships.

I can see patterns in the sizes of the pieces.

- Triangles *a* and *b* cover $\frac{1}{2}$ of the whole so they are each $\frac{1}{4}$ of the square.

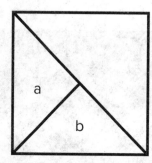

- Triangle *d* covers half of triangle *a*, so triangle *d* is $\frac{1}{8}$ of the whole.

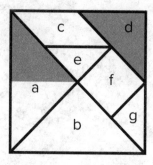

Math is... **Patterns**

What patterns do I see in this figure?

When we do math, we use patterns to solve problems efficiently. Patterns can help you solve problems that are similar.

I can continue to cover the pieces with smaller pieces and determine the fraction of the whole for each piece.

- Triangles *e* and *g* together cover triangle *d*, so together they are also $\frac{1}{8}$ of the whole.
  The two triangles cover square *f*, so square *f* is also $\frac{1}{8}$ of the whole.

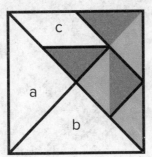

- Parallelogram *c* can also be covered by triangles *e* and *g*, so it must also be $\frac{1}{8}$ of the whole.

By breaking the puzzle into smaller pieces and seeing the relationships among those pieces, I can solve the puzzle.

Work Together

What fraction of the whole do pieces *c* and *e* represent?

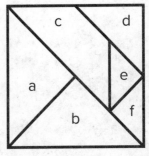

On My Own

Name

The plans for the new park are shown here. What fraction of the park will be used for each type of activity?

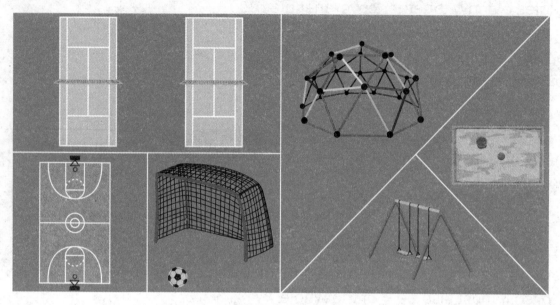

Reflect

How can patterns help you solve problems or equations?

Be Curious

What do you notice?
What do you wonder?

Learn

How do we do math?

▶ When we do math, we often work together.

- We listen carefully.
- We share our thinking.
- We are respectful of others' ideas.
- We critique *the ideas* of others; we don't criticize others.
- We share tools and take turns.

Math is... Mindset

What can I do to be an active listener?

▶ When we do math, sometimes we work on our own.

- We stay focused.
- We look for help when we are stuck.

Math is... Mindset

What can I do to stay focused on my work?

▶ When we do math, we solve problems.

- We make sense of problems.
- We understand the quantities and the relationships among the quantities.
- We don't quit. If we get stuck, we look for different ways.
- We use tools. We select the tool that works best for us.
- We look for patterns.

Math is... Mindset

What can I do when I feel frustrated?

1. What should be the norms when we share our thinking with classmates?

2. What can you do to help all students feel comfortable in math class?

3. How do we use tools appropriately?

4. How can I know when I need help?

On My Own

MATH REPLAY | GO ONLINE

Name

What are two promises our class can make so that we always work together well?

Math is...

 **Reflect**

What are my responsibilities to make sure we can all learn math productively?

Unit Review

Name _____

1. Why is it important to make sense of a problem before solving?

2. How should we respond to our classmates' ideas?

3. How can we decide which tool to use to solve a problem?

4. How can we know when to look for generalizations?

Review

What should be our classroom norms for doing math?
Write up to 5 norms.

1.

2.

3.

4.

5.

Reflect

Choose one of the norms you wrote and tell why it is important.

Fluency Practice   Name ....

Fluency Strategy

You can use partial sums to add whole numbers.

Partial Sums

$$253$$
$$+478$$

11	$3 + 8$
120	$50 + 70$
600	$200 + 400$

731

$$253 + 478 = 731$$

1. Use partial sums to add.

$$465$$
$$+279$$

____	$5 + 9$
____	$60 + 70$
____	$400 + 200$

____

Fluency Flash

Use the base-ten blocks to write equations using partial sums.

2.

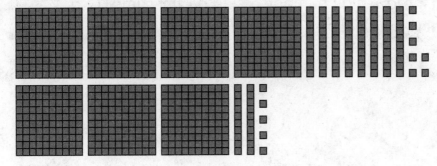

_____

_____

Fluency Check

What is the sum?

3. 621 + 229 = _____

4. 178 + 791 = _____

5. 465 + 517 = _____

6. 513 + 297 = _____

7. 525 + 293 = _____

8. 272 + 458 = _____

9. 128 + 344 = _____

10. 197 + 754 = _____

11. 375 + 426 = _____

12. 362 + 110 = _____

Fluency Talk

How would you explain to a friend how to use partial sums for adding two whole numbers?

How would you add three 2-digit numbers using partial sums? Use an example to explain your thinking.

Generalize Place-Value Structure

Focus Question

How can I use place value to work with multi-digit numbers?

Hi, I'm Poppy.

I want to be a park ranger. Park rangers keep track of animals in the parks, and also the number of visitors. Some parks have hundreds of thousands or even millions of visitors each year.

STEM video | **GO** ONLINE

Name _____

Fewest Coins

Table 1

Use the fewest pennies, nickels, and quarters to make each amount.

Cents	Quarters	Nickels	Pennies
1			
2			
3			
4			
5			
6			
7			
8			
9			
10			
11			
12			
13			
14			
15			
16			
17			
18			
19			
20			
21			
22			
23			
24			
25			
26			

Table 2

Use the fewest pennies, dimes, and dollars to make each amount.

Cents	Dollars	Dimes	Pennies
1			
2			
3			
4			
5			
6			
7			
8			
9			
10			
11			
12			
13			
14			
15			
16			
17			
18			
19			
20			
97			
98			
99			
100			
101			

Understand the Structure of Multi-Digit Numbers

Be Curious

Which doesn't belong?

1,389

180

2,382

11,808

Math is... Mindset

What helps you be motivated to do your best work?

Learn

Akira says that the digits in the number shown are all the same, so they all have the same value.

8,888

Do you agree with Akira's thinking? Explain your reasoning.

Write the number in expanded form. Then look at the value of each part of the expanded form. $$8,888 = 8,000 + 800 + 80 + 8$$ $80 = 8 \times 10$ $800 = 80 \times 10$ $8,000 = 800 \times 10$	The value of each 8 is different. Each 8 represents ten times the value of the 8 to the right. $\times 10 \quad \times 10 \quad \times 10$ 8 , 8 8 8 Akira's thinking is not correct.

The value of a digit is determined by its place-value position.

A digit in one place represents ten times what it represents in the place to its right.

Work Together

How can you describe the relationship between the values of the digits 3 in this number? Explain.

3,830

Name _____

What are the values of the digits in the number?

1. 1,489

    1: _____

    4: _____

    8: _____

    9: _____

2. 98,124

    1: _____

    2: _____

    4: _____

    8: _____

    9: _____

How can you describe the relationship between the values of the underlined digits?

3. 258 and 2,180

4. 16,852 and 14,674

5. 12,184 and 541,247

6. 453 and 1,333

What is the greatest number and the least number you can create using the given digits? Use each digit only once. Do not use 0 as the first digit.

7. 3, 5, 8, and 9

8. 7, 1, 0, 6, 4

9. Is the value of the digit in the hundreds place ten times the value of the digit in the tens place in the number 3,735? Explain.

10. Karma created a number using the digits 4, 2, and 7. Use the following clues to determine Karma's number.

The number is between 7,000 and 8,000.
The digit 4 has a value of 40.
The value of the digit in the thousands place is 10 times the value of the digit to its right.

11. **Extend Your Thinking** Sienna wants to rearrange the digits in the number 1,258,072 so that the value of one of the digits is 10 times the value of another digit in her number. What number could she write? Justify your answer.

12. **Error Analysis** Rahul says the relationship between the 3s in the number 45,339 is different from the relationship between the 6s in the number 66,084. How would you respond to Rahul?

Reflect

How can place value help you determine the value of a digit?

Math is... **Mindset**

How were you motivated to do your best?

Be Curious

What do you notice?
What do you wonder?

Math is... **Mindset**

What can you do to be an active listener?

Learn

How can you read the population of Philadelphia, PA?

Math is... Choosing Tools

What will the tool tell me about the number?

Welcome to
Philadelphia, PA
POP. 1,576,596

You can use a place-value chart to make sense of a multi-digit number.

This place-value chart shows nine positions. It has three groupings. Each grouping is a **period**.

Each period has the same three places.

Millions Period			Thousands Period			Ones Period		
hundreds	tens	ones	hundreds	tens	ones	hundreds	tens	ones
		1	5	7	6	5	9	6

Word Form

One million, five hundred seventy-six thousand, five hundred ninety-six

Standard Form 1,576,596

A comma separates the periods.

Expanded Form 1,000,000 + 500,000 + 70,000 + 6,000 + 500 + 90 + 6

You can use the values of the digits and the names of place-value positions to read and write multi-digit numbers. Commas are used to separate the periods when writing numbers in standard form.

Work Together

How can you write *seven hundred thirty-six thousand, nine hundred two* in standard form and expanded form?

On My Own

Name _____

How can you write the number in standard form?

1. Four hundred thousand, nine hundred thirty _____

2. Thirty-four thousand, nine hundred eighty-nine _____

How can you write the number in expanded form?

3. 530,879

4. 6,216

How can you write the number in word form?

5. 205,782

6. 1,108,308

7. **STEM Connection** Poppy found a sticker on the sign showing the size of Olympic National Park. She knows the size is between one million and nine hundred thousand acres. She also knows that the value of the digit in the ten thousands place is 10 times greater than the value of the digit in the thousands place. What is the size of the park?

Olympic National Park
Established 1938
Size: 🙂2,651 acres

_____

What are other ways to write the number? Complete the table.

	Standard Form	Expanded Form	Word Form
8.	405,832		
9.		500,000 + 30,000 + 9,000 + 10 + 5	
10.			six hundred ten thousand, four hundred sixteen

11. **Extend Your Thinking** How is the word form of 245,007 similar to the word form of 700,245? Explain why these similarities exist.

12. What are the missing words or digits in each form of the number?

Word form: _____ thousand, _____ eight

Expanded form: _____ + 400 + _____

Standard form: 6 _____ , _____ _____ _____

⏱ **Reflect**

How can place value help you make sense of multi-digit numbers?

Math is... **Mindset**

What have you done to be an active listener today?

Be Curious

What question could you ask?

Math is... Mindset

How confident are you that you will be successful today?

Learn

Jonah says that he has walked more steps than Roshni.

How can Jonah support his statement?

Gabe	Roshni	Jonah
10,463 STEPS	10,229 STEPS	10,295 STEPS

Jonah can use place value to compare the two numbers.

Math is... Thinking

What are some mathematical representations you can use to compare numbers?

▶ **One Way** Use a Place-Value Chart

The tens place is the greatest position with different digits.

Thousands Period			Ones Period		
hundreds	tens	ones	hundreds	tens	ones
	1	0	2	9	5
	1	0	2	2	9

9 tens are greater than 2 tens, so 10,295 > 10,229.

▶ **Another Way** Use Expanded Form

$10,295 = 10,000 + 200 + \mathbf{90} + 5$
$10,229 = 10,000 + 200 + \mathbf{20} + 9$

90 is greater than 20, so 10,295 > 10,229.

To compare multi-digit numbers, compare the digits in each place. Start with the digits in the greatest place-value position.

💬 Work Together

Who walked more steps in May? Justify your answer.

Name	Steps in May
Roshni	245,821
Jonah	43,068

On My Own

Name _____

How can you compare the numbers? Complete with >, <, or =.

1. 5,598 ◯ 55,889

2. 123,710 ◯ 123,711

3. 628,910 ◯ 628,800

4. 709,103 ◯ 709,130

5. 6,217 ◯ 6,241

6. 43,829 ◯ 43,598

Is the comparison true or false? Explain your reasoning.

7. 1,780 < 11,780

8. 720,301 < 720,031

9. 34,646 > 321,446

10. 24,747 < 24,774

---

11. Rebecca knows her number is greater than 15,724 by looking at the digits in the tens place. What could be Rebecca's number? Justify your answer.

12. Error Analysis Jamar says 9,280 is greater than 12,621 because the digit 9 is greater than the digit 1. How can you respond to Jamar's statement? Justify your thinking.

13. The table shows the cost of vehicles Maddie's mom is considering. Which vehicle is the most expensive? Justify your answer.

Vehicle	Cost
Minivan	$ 24,990
Pickup Truck	$ 31,990
Sports Car	$ 22,990

14. Extend Your Thinking Write a number less than 4,850 by only switching two digits in this number. Explain your thinking.

🕐 Reflect

How do you describe your answer when you compare numbers?

Math is... Mindset

How did being confident help you succeed?

Round Multi-Digit Numbers

Be Curious

What do you notice?
What do you wonder?

One director said there were about 40,000 visitors at the museum in one month.

Another director said there were about 35,000 visitors at the museum in one month.

Math is... Mindset

How can a different perspective help you with your work today?

Learn

A museum will give a free T-shirt to each visitor in August. The director expects that the number of visitors in August will be about the same as it was in June.

How many T-shirts should the museum director order?

You can round numbers to get a good estimate

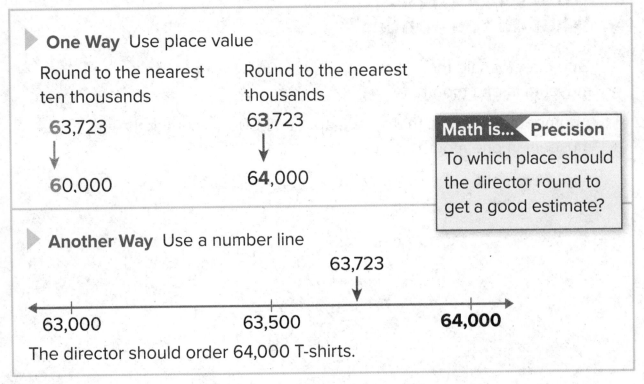

▶ **One Way** Use place value

Round to the nearest ten thousands

63,723

↓

60.000

Round to the nearest thousands

63,723

↓

64,000

Math is... Precision

To which place should the director round to get a good estimate?

▶ **Another Way** Use a number line

63,723

63,000   63,500   **64,000**

The director should order 64,000 T-shirts.

You can round numbers to make estimates. Think about how precise the estimate needs to be when deciding to which place you will round.

Work Together

Each student who participates in field day will get a water bottle, but not all 1,528 students are expected to participate. What is a reasonable estimate of the number of water bottles to order?

On My Own

Name _____

What is your estimate? Round the number as indicated.

1. 478,309 to the
   nearest thousand

2. 105,201 to the nearest
   hundred thousand

3. 95,550 to the nearest
   ten thousand

4. 132,847 to the
   nearest thousand

---

5. **STEM Connection** Denali National Park in Alaska
   has about 650,000 visitors each year. What could
   be the actual number of visitors in one year?
   Explain your reasoning.

---

6. Some astronauts will travel to the moon, which is 238,855 miles
   from the earth.

   a. About how many miles will the astronauts travel there and
      back? Explain the reasoning for your estimate.

   b. How accurate does the estimate need to be?

7. About 15,000 people live in a small town. What could be the actual number of people living in the town?

8. Mae rounds a number to the thousands place and gets 13,000. Eli rounds the same number to the hundreds place and gets 12,600. What is the greatest number that can be rounded as described?

9. Anna and her family will fly 4,387 miles to visit family. What's a reasonable estimate of the distance Anna will fly?

10. A sports stadium has seats for about 100,000 visitors. What could be the actual number of seats?

11. Leon is ordering water bottles for a sports event at which 1,255 people are expected. He plans to round to the nearest thousand to estimate the number of water bottles needed. How would you respond to Leon?

12. Extend Your Thinking Students have collected 1,475 cans for a food drive. Their goal is to collect 2,000 cans. About how many more cans do they need to collect?

⏩ Reflect

How did you apply what you already know about rounding during this lesson?

Math is... **Mindset**

How did looking at things differently help you today?

Rounding Numbers

Name

1. If you round to the nearest hundred, which numbers round to 2,700? Choose all that apply.

   a. 2,752   e. 2,682

   b. 2,735   f. 2,650

   c. 2,749   g. 2,789

   d. 2,599   h. 2,649

   Explain your choices.

---

2. If you round to the nearest hundred, which numbers round to 26,500? Choose all that apply.

   a. 26,449   e. 26,498

   b. 26,385   f. 26,451

   c. 26,589   g. 25,513

   d. 25,389   h. 25,499

   Explain your choices.

3. If you round to the nearest thousand, which numbers round to 26,000? Choose all that apply.

   a. 25,329   **e.** 26,329

   b. 25,781   **f.** 26,585

   c. 25,503   **g.** 26,289

   d. 25,899   **h.** 24,792

Explain your choices.

Reflect On Your Learning

I am confused.

I'm still learning.

I understand.

I can teach someone else.

Unit Review Name _____

Vocabulary Review

Match the word to the phrase that best describes it.

1. round _____
(Lesson 2-4)

A. a way to write a number as a sum that shows the value of each digit

2. period _____
(Lesson 2-2)

B. the form of a number that uses written words

3. expanded form _____
(Lesson 2-1)

C. exactly half the distance between two given numbers

4. standard form _____
(Lesson 2-2)

D. one way to determine a reasonable estimate

5. digit _____
(Lesson 2-1)

E. the usual way of writing a number that shows only its digits

6. word form _____
(Lesson 2-2)

F. a grouping of three digits in numbers

7. halfway point _____
(Lesson 2-4)

G. a symbol used to write numbers

Review

8. What is the relationship between the two 4 digits in the number 904,467? (Lesson 2-1)

9. Which number represents sixty-two thousand, four hundred ninety-five? Choose the correct answer. (Lesson 2-2)

   A. 620,495

   B. 624,95

   C. 62,495

   D. 62,400,095

10. A school raised $8,875. Which shows a reasonable estimate of the amount the school raised? Choose the correct answer. (Lesson 2-4)

   A. $9,500   **B.** $9,000

   C. $7,000   **D.** $8,000

11. What is the value of the digit 2 in 143,287? (Lesson 2-1)

12. What is the word form of 9,284? (Lesson 2-2)

13. In which number does the digit 2 have a value that is ten times the value of the digit 2 in 12,738? Choose the correct answer. (Lesson 2-1)

   A. 26   **B.** 215

   C. 2,387   **D.** 23,901

14. Which of the following are different ways to represent the number 40,381? Choose all that apply. (Lesson 2-2)

   A. $4,000 + 300 + 80 + 1$

   B. Forty thousand, three hundred eighty-one

   C. $40,000 + 300 + 80 + 1$

   D. Four thousand, three hundred eighty-one

   E. $40,000 + 3,000 + 80 + 1$

   F. Forty, three hundred eighty-one

15. What is the value of each digit in the number shown? (Lesson 2-1)

Thousands Period			Ones Period		
hundreds	tens	ones	hundreds	tens	ones
	3	4	4	5	6

16. Michael was playing a video game with his friends. Each person recorded their score in a different form. Who had the highest score? Tell how you know. (Lesson 2-3)

Name	High Score
Paul	three thousand, two hundred fifty-eight
Sameer	2,000 + 900 + 50 + 8
Michael	3,302

17. Which statements are true? Choose all that apply. (Lesson 2-3)

A. 2,315 > 1,319

B. 2,315 < 1,319

C. 1,319 > 2,315

D. 2,315 = 1,319

E. 1,319 < 2,315

18. Keisha has about $3,000 in her savings account. What could be the exact amount in her savings account? Justify your answer.

(Lesson 2-4)

19. Which statements are true? Choose all that apply. (Lesson 2-3)

A. 3,100 = 3,000 + 100

B. 432,238 < 324,239

C. two thousand, six = 2,006

D. 31,840 > 31,440

20. In the number 3,665, how does the value of the digit 6 in the hundreds place compare to the value of the digit 6 in the tens place ? (Lesson 2-1)

21. What is 392,483 rounded to the nearest thousand? (Lesson 2-4)

22. What is 392,483 rounded to the nearest hundred thousand?

(Lesson 2-4)

23. What is the relationship between the two 7 digits in the number 328,277? (Lesson 2-1)

Performance Task

National Park Visitors

There were 642,809 visitors to Denali National Park in Alaska in 2017. In 2015, there were 589,450 visitors to the park.

Part A: What is a reasonable estimate of the number of visitors to Denali National Park in 2017? Explain why it is a reasonable estimate.

Part B: How does the number of visitors to Denali in 2015 compare to the number of visitors in 2017? Write a math statement to represent the comparison. Justify your statement.

Part C: The number of visitors to Everglades National Park in Florida in 2017 was about 600,000. What could be the actual number of visitors to Everglades National Park in 2017? Defend your number.

🕐 Reflect

How does knowing the structure of multi-digit numbers help you work with these numbers?

Fluency Practice Name _____

Fluency Strategy

You can decompose by place value to find the difference.

Decompose

$653 - 212 = ?$
$212 = 200 + 10 + 2$
$653 - 200 = 453$
$453 - 10 = 443$
$443 - 2 = 441$
So, $653 - 212 = 441$

1. How can you decompose by place value to find the difference?

$697 - 324 = ?$

$324 = $ _____ $+$ ____ $+$ ___

$697 - $ _____ $= 397$

$397 - $ ____ $= 377$

$377 - $ __ $= $ _____

Fluency Flash
Use the base-ten blocks to write matching subtraction equations.

2.

_____

3.

_____

Fluency Check

What is the sum or difference?

4. $739 - 428 = ?$ _____

5. $238 + 684 = ?$ _____

6. $723 + 246 = ?$ _____

7. $736 + 125 = ?$ _____

8. $858 - 615 = ?$ _____

9. $958 - 230 = ?$ _____

10. $684 - 152 = ?$ _____

11. $549 + 287 = ?$ _____

12. $164 + 528 = ?$ _____

13. $356 - 145 = ?$ _____

14. $674 - 213 = ?$ _____

15. $464 + 103 = ?$ _____

Fluency Talk

How would you explain to a friend how to decompose a number by place value to make subtraction easier?

How is using partial sums to add like decomposing a number by place value to subtract?

Addition and Subtraction Strategies and Algorithms

Focus Question

How can I add and subtract with strategies and algorithms?

Hi, I'm Hiro.

I want to be an ocean engineer and I'm helping out at the rescue center. I need to find the total weight of all the sea lions and harbor seals at the center. Multi-digit addition and subtraction help me do my job!

Name

The Greatest Sum or Difference

Listen to your teacher. Make the greatest sum or difference.

A.

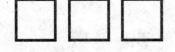

\+ ⬜⬜⬜

B.

⬜⬜⬜

− ⬜⬜⬜

C.

⬜⬜⬜

\+ ⬜⬜⬜

D.

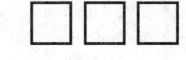

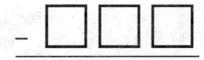

− ⬜⬜⬜

E.

⬜⬜⬜⬜

\+ ⬜⬜⬜⬜

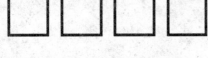

F.

⬜⬜⬜⬜

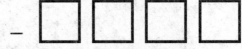

− ⬜⬜⬜⬜

G.

⬜⬜⬜⬜

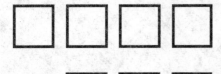

\+ ⬜⬜⬜

H.

⬜⬜⬜⬜

− ⬜⬜⬜

Estimate Sums or Differences

Be Curious

What's the question?

A school principal has some money to spend on new playground equipment.

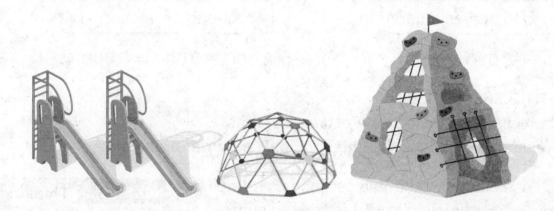

Math is... **Mindset**

How can you show others that you value their ideas?

Learn

A school principal has about $8,000 to spend on new playground equipment. The table shows the top three items students selected.

Can the principal order the double slide and the climbing wall?

Playground Equipment	
Item	**Cost**
Double Slide	$3,919
Geodome	$2,218
Climbing wall	$3,564

You can estimate the total cost of the two items.

> **One Way** Round to the nearest hundred.
>
> 3,919 + 3,564 = ?
>
> | Round to nearest hundred |
>
> 3,900 + 3,600 = 7,500

> **Another Way** Use **front-end estimation.**
>
> 3,919 + 3,564 = ?
>
> | Front-end estimation |
>
> 3,000 + 3,000 = 6,000

3,919 + 3,564 = 7,483
The exact sum is close to the estimated sums.
It is reasonable to think that the principal can order both items.

You can use estimation strategies, such as rounding and front-end estimation to find estimated sums. Estimating sums or differences is useful to determine the reasonableness of a calculated solution.

> **Math is... Thinking**
>
> Why is it important to consider the situation when choosing an estimation strategy?

Work Together

About how much more money will the school principal need to purchase all three items? Explain your thinking.

On My Own

Name _____

How can you estimate the sum or difference? Explain your strategy.

1. $12{,}258 + 14{,}926 =$ _____

2. $5{,}246 - 392 =$ _____

How can you estimate the sum or difference? Use a calculator to find the actual answer. Circle the estimate closest to the actual sum or difference.

	Rounding	Front-end estimation
3. $8{,}303 - 2{,}789 = ?$		
4. $3{,}783 + 1{,}416 = ?$		
5. $3{,}155 + 2{,}205 = ?$		
6. $9{,}875 - 4{,}968 = ?$		
7. $4{,}228 + 986 = ?$		

8. How can you estimate the sum of $2{,}352 + 8{,}761$? Explain your strategy.

9. Anton wrote the equation below. Is the difference reasonable? Explain your thinking.

$$1{,}988 - 713 = 275$$

10. Springfield School District had 1,578 students last year. This year it has 2,138 students. Carmen says that the number of students increased by over 500 students. Is her statement reasonable?

11. STEM Connection Hiro helped calculate the weight of two northern fur seal pups. He said the total weight of the two pups was over 15,000 grams. Is Hiro's estimate reasonable? Explain.

Pup 1	Pup 2
8,250 grams	7,920 grams

12. Extend Your Thinking Tanya walked 9,526 steps. Her brother Marcus walked 7,488 steps. Tanya says that she walked about 3,000 more steps than Marcus. Marcus says that the difference is closer to 2,000 steps. Whose estimate do you agree with? Explain why.

⟳ Reflect

What do you need to consider when choosing an estimation strategy?

Math is... **Mindset**

How have you shown others that you value their ideas?

Unit 3
Estimation

Name _____

Four students showed their work to estimate this sum:

$$547 + 231 + 363$$

Decide if each student's process provides a correct way to estimate the sum.

Student A	Explain why you chose Yes or No.
I added: 500 + 200 + 400. *My estimate is 1,100.* Circle Yes or No. **Yes No**	

Student B	Explain why you chose Yes or No.
First I added the numbers. $547 + 231 + 363 = 1{,}141$ *Then I rounded. My estimate is 1,140.* Circle Yes or No. **Yes No**	

Four students showed their work to estimate this sum:

$$547 + 231 + 363$$

Decide if each student's process provides a correct way to estimate the sum.

Student C	Explain why you chose Yes or No.
I found this sum: 550 + 225 + 375. *My estimate is 1,150.* Circle Yes or No. **Yes** **No**	

Student D	Explain why you chose Yes or No.
I found three sums, decomposing *the hundreds, tens, and ones. Then* *I added those sums:* *500 + 200 + 300 = 1,000* *40 + 30 + 60 = 130* *7 + 1 + 3 = 11* *My estimate: 1,141* Circle Yes or No. **Yes** **No**	

Reflect On Your Learning

I am
confused.   I'm still
learning.   I understand.   I can teach
someone else.

Strategies to Add Multi-Digit Numbers

Be Curious

What do you notice?
What do you wonder?

1,613 pounds

1,297 pounds

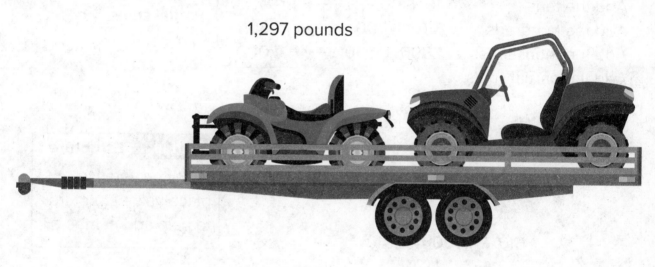

Math is... Mindset

What can you do today to help build a relationship with a classmate?

Learn

Vashti has a trailer with a weight limit of 3,000 pounds.

Can both vehicles be safely loaded on the trailer?

1,297 pounds   1,613 pounds

You can add to determine the total weight of the two vehicles.

▶ **One Way** Add using **partial sums**.

$$
\begin{array}{r}
1,613 \\
+\ 1,297 \\
\end{array}
$$

Add the ones	$3 + 7$	$\longrightarrow$	10 ⎫
Add the tens	$10 + 90$	$\longrightarrow$	100 ⎪ Partial sums
Add the hundreds	$600 + 200$	$\longrightarrow$	800 ⎬
Add the thousands	$1,000 + 1,000$	$\longrightarrow$	2,000 ⎭
Add the partial sums.			2,910

▶ **Another Way** Add by adjusting the numbers.

$1,613 + 1,297$

-3 $+3$

$1,610 + 1,300 = 2,910$

> **Math is... Structure**
>
> How can you use place-value structure to help add numbers?

Both vehicles weigh 2,910 pounds, so they can be safely loaded on the trailer.

You can use addition strategies that you know to add multi-digit numbers.

🗨 Work Together

The trailer that holds the vehicles weighs 1,550 pounds. What is the total weight of the vehicles and the trailer? Explain the strategy you used to solve.

On My Own

Name _____

What is the sum?

1. $2,582 + 493 =$ _____

2. $476 + 8,719 =$ _____

3. $1,945 + 3,289 =$ _____

4. $12,017 + 5,308 =$ _____

5.
$$\begin{array}{r} 26,118 \\ +\ 11,043 \\ \hline \end{array}$$

6.
$$\begin{array}{r} 47,621 \\ +\ 21,345 \\ \hline \end{array}$$

7.
$$\begin{array}{r} 101,253 \\ +\ 27,285 \\ \hline \end{array}$$

8. The indoor water park had 10,242 visitors in January and 11,495 visitors in February. What was the total attendance for the two months?

9. **Extend Your Thinking** The book bank collected 13,962 books last year. This year it collected 15,185 books. The book bank expects to collect about the same number of books next year as it did this year. About how many books will be collected all three years? Explain your answer.

10. How can you add 11,864 + 9,599 by adjusting the addends? Show your strategy.

11. **Error Analysis** Macy completed the problem below. How can you help Macy understand her error and find the correct sum?

$$
\begin{array}{r}
5,331 \\
+\ 2,702 \\
\hline
3 \\
30 \\
100 \\
7,000 \\
\hline
7,133
\end{array}
$$

12. **STEM Connection** The weight of a Stellar Sea Lion and a California Sea Lion are shown in the table. What is the total weight of the two sea lions?

Animal	Weight (kilograms)
Stellar Sea Lion	1,026
California Sea Lion	395

Reflect

How did you decide which strategy to use to find each sum?

Math is... **Mindset**

What have you done today to help build a relationship with a classmate?

Understand an Addition Algorithm

Be Curious

**How are they the same?
How are they different?**

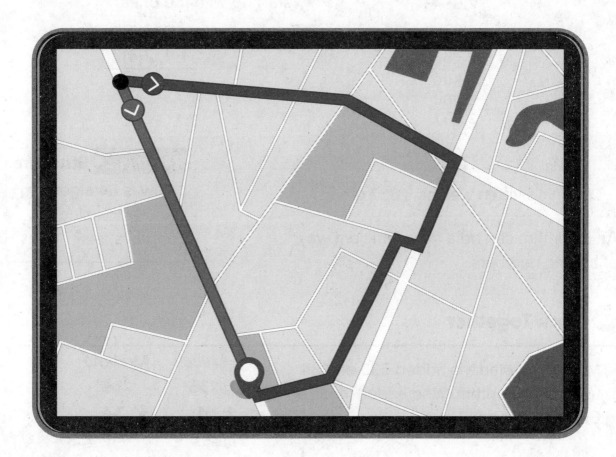

Math is... Mindset

What do you want your classmates
to know about your math story?

Learn

November

The weight of Dominic's kitten is now 1,243 grams more than it was in November.

What is the weight of Dominic's kitten now?

You can add to solve the problem. Write the addends vertically, aligning the digits by place value, to find the sum.

2,654 grams

▶ **One Way** Use partial sums. Add by place value.

```
    2,654
  + 1,243
        7
       90
      800
    3,000
    3,897
```

▶ **Another Way** Use an **algorithm**. Add by place value starting with the ones.

```
  2,6 5 4
+ 1,2 4 3
  3,8 9 7
```

Dominic's kitten weighs 3,897 grams.

An algorithm can be a more efficient way of adding numbers.

> **Math is...** **Structure**
>
> How is an algorithm similar to using partial sums?

⊙ Work Together

Mabel and Maderia added 235 and 34 using an algorithm. Who added incorrectly? Explain the error.

Mabel
```
  235
+  34
  575
```

Maderia
```
  235
+  34
  269
```

On My Own

Name _____

What is the sum? Use an algorithm to solve.

1. 4,380
    + 612
    _____

2. 12,943
    + 4,036
    _____

3. 42,818
    + 7,120
    _____

4. 8,405
    + 1,571
    _____

5. 7,364
    + 2,321
    _____

6. 4,129
    + 2,530
    _____

7. A business purchased a copier for $1,217 and a laptop for $761. How much did the business spend on both items? Use an algorithm to solve.

8. A factory made 64,457 car parts in the first three weeks of the month and 3,502 car parts in the fourth week of the month. How many car parts did the factory make in the four weeks?

9. Aria is calculating 64,203 + 23,562 by using partial sums. Show what her work could look like. Then complete the equation.

64,203 + 23,562 = _____

10. A band played two concerts with a total attendance of 9,698 people. The first concert had 4,467 people in attendance. How many people attended the second concert? Write an addition equation to solve.

11. STEM Connection A ship studying marine animal populations traveled 1,183 miles on the first part of its mission. The ship will travel another 815 miles to complete its mission. What is the total distance the ship will travel on this mission? Explain how you knew what math operation to use to solve the problem.

12. Extend Your Thinking Add 3,616 and 5,372 using partial sums and an algorithm. Explain how the two methods are similar and different.

 **Reflect**

How does place value help you add using an algorithm?

> **Math is... Mindset**
>
> How have you helped your classmates know about your math story?

Understand an Addition Algorithm Involving Regrouping

Be Curious

What do you notice?
What do you wonder?

15,852
MILES

8,849
MILES

Math is... **Mindset**

What makes you feel confidence about your work today?

Learn

The activity tracker shows the number of steps Marta has before taking a hike. During the hike, Marta's tracker records another 8,238 steps.

Marta

14,156 STEPS

How many steps does Marta have after the hike?

You can add using an algorithm to determine Marta's steps.

Step 1 Add the ones.

$6 + 8 = 14$

Regroup 14 as 1 ten and 4 ones.

$$\begin{array}{r} \overset{1}{1}4,1\overset{1}{5}6 \\ +\ 8,238 \\ \hline 22,394 \end{array}$$

Step 2 Add the tens.

$10 + 50 + 30 = 90$

Step 3 Add the hundreds.

$100 + 200 = 300$

> **Math is...** Explaining
>
> How can you justify using an algorithm to solve multi-digit addition problems?

Step 4 Add the thousands.

$4,000 + 8,000 = 12,000$

Regroup 12,000 as 1 ten thousand and 2 thousands.

Step 5 Add the ten thousands.

$10,000 + 10,000 = 20,000$

Marta has 22,394 steps after the hike.

The algorithm has a process for recording sums of numbers that require regrouping.

🗨 Work Together

Kaden walks 6,972 steps on Monday. Tuesday, he walks 6,344 steps. How many steps did Kaden walk on Monday and Tuesday? Use an algorithm to solve.

On My Own

Name _____

What is the sum? Use an algorithm to solve.

1. 1,458
    + 926
    ‾‾‾‾‾‾‾‾

2. 4,239
    + 765
    ‾‾‾‾‾‾‾‾

3. 2,744
    + 1,306
    ‾‾‾‾‾‾‾‾

4. 4,827
    + 3,505
    ‾‾‾‾‾‾‾‾

5. 9,087
    + 7,668
    ‾‾‾‾‾‾‾‾

6. 12,058
    + 4,867
    ‾‾‾‾‾‾‾‾

7. A car manufacturer has made 4,569 cars so far this month. They will make 5,286 more cars this month. How many cars will they make this month? Use an algorithm to solve.

8. Trevon had 1,425 trading cards in his collection. He traded many cards and now has 395 more cards than he started with. How many trading cards does he have now? Use an algorithm to solve.

9. Luca and his family are taking a road trip.

   a. The first day they drove from Chicago to Omaha, 467 miles. The next day they drove from Omaha to Billings, 838 miles. How many miles did they drive the first two days?

   b. Luca's family then drove to Salt Lake City and Los Angeles, a total of 1,238 miles. How many miles have they driven so far on their trip?

   c. Luca and his family ended their trip in Los Angeles and drove back to Chicago a different way. When they stopped exactly halfway through the trip home, they had driven 1,259 miles. How many miles did they travel on the entire trip home?

10. Extend Your Thinking Fill in the missing digits. Explain how you found each digit.

$$
\begin{array}{r}
2\ \square\,,\ 1\ \square\ 9 \\
+\qquad 8\ 4\ \square \\
\hline
2\ 2\,,\ \square\ 0\ 4
\end{array}
$$

Reflect

How is using an algorithm with regrouping different from using the algorithm without regrouping? How is it the same?

Math is... **Mindset**

What helped you feel confidence about your work today?

Strategies to Subtract Multi-Digit Numbers

Be Curious

What do you notice?
What do you wonder?

VOTING RESULTS

9,856   6,298   3,588   1,151

Math is... **Mindset**

What are some ways you can connect with your classmates?

Learn

In an election for mayor with two candidates, 9,856 people voted.

How many votes did the other candidate receive?

★★ **VOTING** ★★
★★ **RESULTS** ★★

New Mayor elected with 6,298 votes.

You can subtract to determine the votes the other candidate received.

▶ **One Way** Use place value to subtract.

Decompose the number you are subtracting. Then subtract the parts.

$6,298 = 6,000 + 200 + 90 + 8$

$9,856 - 6,000 = 3,856$

$3,856 - 200 = 3,656$

$3,656 - 90 = 3,566$

$3,566 - 8 = 3,558$

The other candidate received 3,558 votes.

▶ **Another Way** Adjust the numbers to make them easier to subtract.

$9,856 - 6,298 = ?$

+2   +2

$9,858 - 6,300 = 3,558$

Math is... Explaining

What do you need to consider when adjusting numbers to subtract?

You can use strategies you know to make it easier to subtract multi-digit numbers.

💬 Work Together

In a county election, 286,787 people voted. The winning candidate received 72,455 votes. How many votes did the other candidates receive?

On My Own

Name _____

How can you decompose to subtract? Find the difference.

1. $2,532 - 1,301 = $ _____

2. $6,489 - 2,472 = $ _____

3. $8,018 - 7,659 = $ _____

4. $11,023 - 1,414 = $ _____

How can you adjust to subtract? Find the difference.

5. $12,469 - 10,212 = $ _____

6. $97,137 - 24,677 = $ _____

7. $46,597 - 4,267 = $ _____

8. $84,649 - 126 = $ _____

9. A restaurant served 14,299 meals in January and 13,039 meals in February. How many more meals did the restaurant serve in January than in February?

10. The first night of a play 3,568 tickets were sold. The second night 2,984 tickets were sold. How many more tickets were sold on the first night?

11. **Extend Your Thinking** What two different strategies can you use to find the difference? How are the two strategies similar? How are they different?

$$15,736 - 10,302 = \underline{\hspace{1.5cm}}$$

12. **Error Analysis** Rafael and Sadia solved a problem by adjusting. Which student solved correctly? Explain your answer.

Rafael: $9,798 - 6,098 = ?$

$+2$ -2

$9,800 - 6,096 = 3,704$

Sadia: $9,798 - 6,098 = ?$

$+2$ $+2$

$9,800 - 6,100 = 3,700$

Reflect

How did you decide which strategy to use to find each difference?

Math is... **Mindset**

How have you connected with your classmates?

?

Be Curious

What question could you ask?

African Elephant
5,386 kilograms

Asian Elephant
4,185 kilograms

Math is... Mindset

What goal do you want to
accomplish today?

Learn

What is the difference in weight between the African elephant and the Asian elephant?

African Elephant
5,386 kilograms

Asian Elephant
4,185 kilograms

You can use an algorithm to determine the difference in weight.

Subtract by place value starting on the right.

Step 1   Subtract the ones.
$6 - 5 = 1$

Step 2   Subtract the tens.
$80 - 80 = 0$

$$\begin{array}{r} 5,386 \\ -\,4,185 \\ \hline 1,201 \end{array}$$

Step 3   Subtract the hundreds.
$300 - 100 = 200$

Step 4   Subtract the thousands.
$5,000 - 4,000 = 1,000$

> **Math is...** ◀ **Structure**
>
> How can you apply what you know about the addition algorithm to use the subtraction algorithm?

The African elephant weighs 1,201 kilograms more than the Asian elephant.

When you use an algorithm to subtract, align the numbers vertically by place value and then subtract the digits starting at the ones places.

⊙ Work Together

A gray whale weighs 68,728 pounds. A whale shark weighs 43,203 pounds. What is the difference in their weights?

On My Own

Name _____

What is the difference? Use an algorithm to solve.

1. 1,558
    −   247
    _____

2. 53,720
    − 33,400
    _____

3. 4,964
    − 2,803
    _____

4. 48,579
    −   4,222
    _____

5. 12,923
    − 10,712
    _____

6. 2,646
    − 1,335
    _____

7. 7,438
    − 5,225
    _____

8. 267,982
    − 132,580
    _____

9. Addie and her family are driving to Florida to see her grandmother. The trip is 1,387 miles. They drove 365 miles the first day. How many miles do they have left to drive?

10. A summer camp is building new cabins. They spent $2,789 for wood and tools to complete the project. The tools cost $1,024. How much did the summer camp spend on wood?

11. Fill in the missing digits. Explain how you found each digit.

$$
\begin{array}{r}
2,\ 7\ 4\ \square \\
-\ 1,\ \square\ \square\ 4 \\
\hline
\square,\ 5\ 4\ 1
\end{array}
$$

12. Extend Your Thinking Starwood School has an annual walkathon. Was the increase greater from two years ago to last year, or from last year to this year? Explain how you know.

Walkathon Fundraiser		
This Year	**Last Year**	**2 Years Ago**
$7,875	$5,652	$3,420

Reflect

Why is it helpful to write numbers vertically when you use the subtraction algorithm?

Math is... **Mindset**

How have you worked to accomplish your goal today?

Understand a Subtraction Algorithm Involving Regrouping

Be Curious

How are they the same?
How are they different?

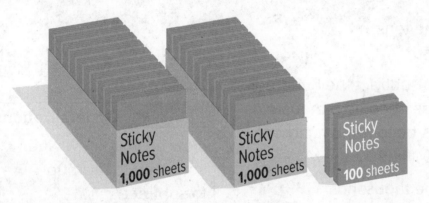

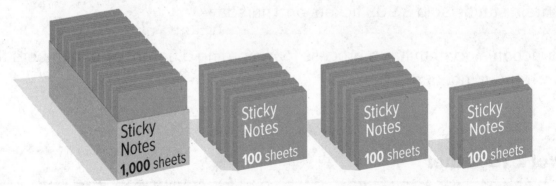

Math is... Mindset

How can you show you understand how others are feeling?

Learn

The number of tickets sold at the train station on Wednesday is shown. On Thursday, 2,475 fewer tickets were sold.

How many tickets were sold on Thursday?

Train Tickets

SOLD 6,384

You can use an algorithm to determine the number of tickets sold.

Step 1 Regroup to subtract the ones.
8 tens become 7 tens.
4 ones become 14 ones.
$14 - 5 = 9$

$$\begin{array}{r} \overset{5}{\cancel{6}},\overset{13}{\cancel{3}}\overset{7}{\cancel{8}}\overset{14}{\cancel{4}} \\ -\,2,475 \\ \hline 3,909 \end{array}$$

Step 2 Subtract the tens.
$70 - 70 = 0$

Step 3 Regroup to subtract the hundreds.
6 thousands become 5 thousands.
3 hundreds become 13 hundreds.
$1,300 - 400 = 900$

> **Math is... Generalizations**
> How do you know when you need to regroup to subtract?

Step 4 Subtract the thousands.
$5,000 - 2,000 = 3,000$

The train station sold 3,909 tickets on Thursday.

A subtraction algorithm has a process for recording differences of numbers that require regrouping.

Work Together

The train station sold 7,025 tickets on Friday. On Saturday, 1,249 fewer tickets were sold than on Friday. How many tickets were sold on Saturday?

On My Own

Name _____

What is the difference? Solve using an algorithm.

1. $\begin{array}{r} 7,570 \\ -453 \\ \hline \end{array}$

2. $\begin{array}{r} 33,071 \\ -2,893 \\ \hline \end{array}$

3. $\begin{array}{r} 12,050 \\ -7,983 \\ \hline \end{array}$

4. $\begin{array}{r} 4,382 \\ -633 \\ \hline \end{array}$

5. $\begin{array}{r} -67,821 \\ 7,954 \\ \hline \end{array}$

6. $\begin{array}{r} 172,005 \\ -48,273 \\ \hline \end{array}$

7. $\begin{array}{r} 6,805 \\ -4,782 \\ \hline \end{array}$

8. $\begin{array}{r} 87,034 \\ -77,245 \\ \hline \end{array}$

Solve. Show your work.

9. A baseball league had 41,384 hits in the fall season and 42,215 hits the spring season. How many more hits did the league have in the spring season?

10. A company bought a delivery truck for $35,698. A year later the company bought a second delivery truck for $39,105. How much more did they spend on the second truck?

11. **Error Analysis** Julio subtracted using an algorithm. Do you agree or disagree with his solution? Explain.

$$
\begin{array}{r}
22,7\overset{11}{\cancel{2}}\overset{13}{\cancel{3}} \\
-12,644 \\
\hline
10,179
\end{array}
$$

12. **Extend Your Thinking** A new video game was released and sold 263,498 copies in the first month. In the second month after the release, 12,498 fewer games were sold. In the third month after the release, 120,512 games were sold. What is the difference between the number of games sold in the second and third months?

🐾 Reflect

How is the subtraction algorithm with regrouping similar to the addition algorithm with regrouping? How is it different?

> **Math is...** **Mindset**
> How have you shown you understand how others are feeling?

Represent and Solve Multi-Step Problems

Be Curious

What's the question?

Tori played the first three levels of a video game.

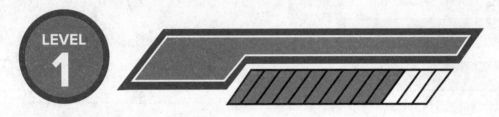

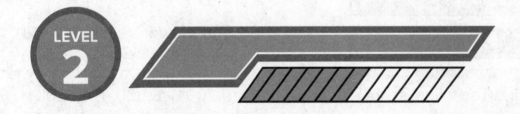

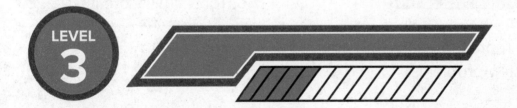

Math is... **Mindset**

What helps you know when there is a problem?

Learn

Tori played the first three levels of a video game.

- She scored 9,200 points in Level 1.

- She scored 2,700 fewer points in Level 2 than Level 1.

- She scored 1,980 fewer points in Level 3 than Level 2.

What was Tori's score after 3 levels?

You can solve the problem in steps.

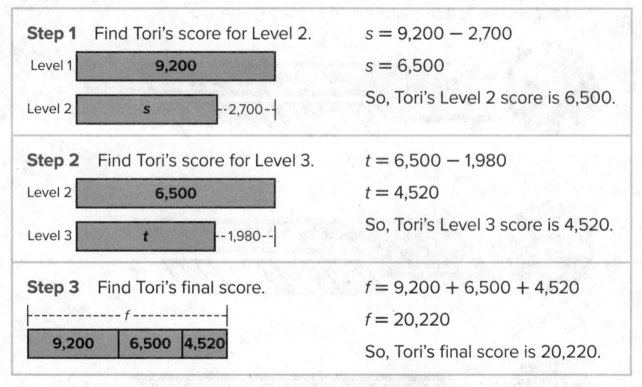

Step 1 Find Tori's score for Level 2.

| Level 1 | 9,200 |
| Level 2 | s --2,700-- |

$s = 9,200 - 2,700$

$s = 6,500$

So, Tori's Level 2 score is 6,500.

Step 2 Find Tori's score for Level 3.

| Level 2 | 6,500 |
| Level 3 | t --1,980-- |

$t = 6,500 - 1,980$

$t = 4,520$

So, Tori's Level 3 score is 4,520.

Step 3 Find Tori's final score.

| f |
| 9,200 | 6,500 | 4,520 |

$f = 9,200 + 6,500 + 4,520$

$f = 20,220$

So, Tori's final score is 20,220.

You can use representations and equations with **variables** to solve multi-step problems.

⊙ Work Together

There were some people at the soccer stadium when the game started. Then, another 1,120 people arrived. By halftime another 8,456 people had arrived. Then, 2,011 people left. Now there are 13,807 people in the stadium. How many people were in the stadium when the game started?

On My Own

Name _____

Use diagrams and equations with variables to solve the problem.

1. Jamar needs sequins for costumes for a school play. The king's costume needs 3,250 sequins. The queen's costume needs 1,750 more sequins than the king's costume. The jester's costume needs 750 fewer sequins than the queen's costume. How many sequins does Jamar need for all three costumes?

2. There are 550 students eating lunch in four different picnic areas of the zoo. How many students are eating lunch at Flamingo Feast?

Picnic Area	Number of Students
Giraffe Jump	217
Manatee Munch	138
Gorilla Garden	97
Flamingo Feast	?

3. An art teacher had 140 jars of paint. In the first half of the year, her students used 95 jars of paint. The teacher bought 35 more jars of paint. At the end of the year, she had 15 unused jars of paint. How many jars of paint did her students use in the second half of the year?

4. The cafeteria distributed 940 cartons of milk at breakfast and 1,670 cartons of milk at lunch. The cafeteria had 7,036 cartons of milk at the end of the day. How many cartons of milk did the cafeteria have at the beginning of the day?

5. **STEM Connection** An ocean engineer used a sonar system to count fish populations of four schools of fish.

School	Number of Fish
1	234
2	536
3	1,112
4	189

a. How many fish were counted?

b. The sonar used to count the fish is not always accurate, it can have an error of counting 10 fish too many or 10 fish too few for every 1,000 fish. Would it be reasonable to say that there could be 2,090 fish in all four schools? Explain.

6. **Extend Your Thinking** Gina used the following equations to solve a multi-step problem. Write a word problem that can be represented by the equations.

Step 1: 5,700 − 2,200 = 3,500

Step 2: 3,500 − 1,545 = 1,955

Step 3: 5,700 + 3,500 + 1,955 = 11,155

⊘ Reflect

How do you decide whether to write an addition equation or a subtraction equation for each step in a problem?

Math is... Mindset

How have you known when there is a problem?

Solve Multi-Step Problems Involving Addition and Subtraction

Be Curious

What math do you see in this problem?

The Save My Forest group had some money in their account. They spent some money on building new trails. They now need some money for cleaning up the lake and some money for conservation training.

Math is... Mindset

What helps you feel calm when you feel angry?

Learn

The Save My Forest group had $268,495 in their account. They spent $129,390 on building new trails. They now need $86,200 for cleaning up the lake and $63,750 for conservation training.

How much more money does the group need for cleanup and training?

You can solve the problem in steps.

Step 1   Find the amount the group has left after building the trails.

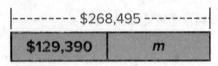

$m = 268,495 - 129,390$
$m = 139,105$

The group has $139,105 left.

Step 2   Find the cost of the cleanup and training.

$86,200 + 63,750 = 149,950$

The cleanup and training will cost $149,950.

Step 3   Find how much more money the group needs.

$149,950 - 139,105 = 10,845$

The group needs $10,845 more for the cleanup and training.

> **Math is...** **Exploring**
> What other strategies could have been used?

Solving problems sometimes involves using several steps and multiple strategies.

⟲ Work Together

The Save Forest Tigers group had $90,357 in their account. They spent $74,240 on rescue efforts during the year. They raised $41,500 by having a fundraiser. The group then spent $19,450 on an awareness campaign. How much is in their account?

On My Own

Name _____

Solve. Show your work.

1. Michael and his family are driving 2,120 miles across the country over four days. The first day, they drove 517 miles. The second day, they drove 535 miles. The third day, they drove 493 miles. How many miles do they have left to drive?

2. A school is collecting cans for the local food pantry. The first week they collected 1,415 cans. The second week they collected 973 cans. The third week they collected 200 cans more than they collected in the second week. After the fourth week, they had a total of 4,542 cans. How many cans did they collect the fourth week?

3. The Honey Bears Stadium sold 25,050 student tickets and 30,975 general admission tickets for their first game. For their next game they sold 27,365 student tickets and 18,527 general admission tickets. How many more tickets did they sell for the first game?

4. The Town Hall Players were selling tickets to their weekend show. They sold 5,789 tickets in advance. At the door, they sold 512 tickets on Friday and 928 on Sunday. They sold a total of 8,054 tickets. How many tickets did they sell at the door on Saturday?

5. **STEM Connection** A northern fur seal pup was rescued and weighed 7,125 grams. The seal's weight increased by 3,238 grams in the first month and 4,615 grams in the second month. The seal needs to weigh at least 18,000 grams in order to be released back into the ocean. Does the seal weigh enough to be released? If not, explain how you can determine how much weight the seal needs to gain before he can be released.

6. **Extend Your Thinking** A garden club wants to purchase land that costs $18,287. In January, before collecting monthly dues, the club has $15,185 in its account. The club collects $500 in membership dues each month. In what month will the garden club be able to purchase the land? Explain.

🕐 **Reflect**

Choose a problem you have solved. Explain how you could solve the problem in another way.

Math is... Mindset

How have you stayed calm when you feel angry?

Unit Review Name _____

Vocabulary Review

Choose the correct word(s) to complete the sentence.

algorithm	compatible numbers
front-end estimation	partial sums
regroup	variable

1. A(n) _____ is a letter that represents an unknown quantity. (Lesson 3-8)

2. A(n) _____ is a way of doing something in math. It is a set of steps that, if done correctly, always works. (Lesson 3-3)

3. _____ are numbers in a problem that are easy to work with mentally. (Lesson 3-2)

4. When using _____, you keep the front, or leftmost, digit and all other digits become zeros. (Lesson 3-1)

5. You can use _____ to add numbers. Add one place value at a time, and then add those sums to find the total sum. (Lesson 3-3)

6. To _____ means to use place value to exchange equal amounts when renaming a number. (Lesson 3-4)

Review

7. A company logs the number of minutes employees use on company cell phones. In April, the employees used 2,843 minutes. In May, the employees used 4,923 minutes. Which is a good estimate of the number of minutes the employees used in April and May? (Lesson 3-1)

   A. 5,000 minutes

   B. 6,000 minutes

   C. 8,000 minutes

   D. 9,000 minutes

8. An office building sells for $350,000. A house nearby sells for $245,600. How much more money does the office building sell for? (Lessons 3-1, 3-6)

9. A concession stand has $620. They earn $525 selling popcorn and $162 selling drinks. They spend $365 to order more supplies. Write an equation to show how much money the concession stand has now.

(Lesson 3-8)

10. An elementary school is collecting box tops. The table shows how many box tops each grade has collected.

Grade	Box Tops
1	2,475
2	3,256
3	1,982
4	4,034

Which two grades combined have collected about 5,000 box tops? Explain how you found your answer. (Lesson 3-1)

11. A bakery sold 1,353 muffins on Friday and 2,231 muffins on Saturday. How many muffins did the bakery sell? (Lessons 3-2; 3-3)

12. Find 39,491 + 56,356. (Lesson 3-4)

13. Find 6,763 − 5,498. (Lesson 3-5)

14. Find 71,395 − 6,627. (Lesson 3-7)

15. Find 10,252 + 1,298. (Lesson 3-4)

16. A theatre has 1,485 tickets to sell. They sell 691 tickets for main floor seats and 478 tickets for balcony seats. There are 184 tickets still available for balcony seats. How many tickets are still available for main floor seats? Explain how you found your answer. (Lesson 3-9)

17. Shandra wins 2,184 tickets at the arcade. Mavis wins 1,135 fewer tickets than Shandra. How many tickets does Mavis win? (Lesson 3-7)

18. A summer camp has an art program and a music program. The art program has 1,053 more students than the music program. If the music program has 2,118 students, how many students are in the art program?

(Lessons 3-2, 3-4)

19. There are 551 cars parked in an airport parking garage. The first level has 128 empty spaces. The second level has 294 empty spaces. The third level has 302 empty spaces. How many spaces are in the parking garage? Explain how you found your answer. (Lesson 3-9)

Performance Task

Ocean engineers help develop systems that make cargo ships more efficient when traveling on the ocean. A cargo ship can be loaded with thousands of shipping containers at one time. The Shining Star cargo ship is loaded with 1,112 shipping containers to deliver to three stops.

Part A: After the first stop, there are 1,008 shipping containers left on the ship. How many shipping containers are unloaded at the first stop? Write and solve an equation to represent the problem.

Part B: More shipping containers are unloaded from the ship at the second stop. At the third stop, the remaining 878 shipping containers will be unloaded. How many shipping containers are unloaded at the second stop? Write and solve an equation to represent the problem.

Part C: At the third stop, there are already 13,567 shipping containers. How many pieces of cargo will be at the stop after the ship is unloaded? Write and solve an equation to represent the problem.

Reflect

Describe two different ways you can add and subtract multi-digit numbers.

Fluency Practice Name _____

Fluency Strategy

You can use strategies and patterns in products to multiply fluently.

You can use doubling to multiply by 2.

$2 \times 6 = ?$
Double 6 to find the product. $6 + 6 = 12$
So, $2 \times 6 = 12$

You can use patterns in products of 5 and 10 to multiply fluently.

$5 \times 5 = 25$ $5 \times 6 = 30$
Products of 5 have a 0 or 5 in the ones place.
$10 \times 4 = 40$
Products of 10 have a 0 in the ones place.

You can use a 10s fact to find a 5s fact.

$5 \times 3 = ?$
$10 \times 3 = 30$. 5 is half of 10, so 5×3 is half of 30.
$5 \times 3 = 15$

1. What is 8×10? 10 is double _____, and $8 \times$ _____ = _____

So, $8 \times 10 =$ _____

Fluency Flash

Write an addition and a multiplication equation for the model.

2.

3.

_____ + _____ = _____   _____ + _____ = _____

_____ × _____ = _____   _____ × _____ = _____

Fluency Check

What is the sum, difference, or product?

4. $7 \times 10 =$ _____

5. $496 + 327 =$ _____

6. $9 \times 5 =$ _____

7. $636 + 182 =$ _____

8. $2 \times 4 =$ _____

9. $926 - 547 =$ _____

10. $5 \times 7 =$ _____

11. $4 \times 5 =$ _____

12. $526 + 389 =$ _____

13. $687 - 243 =$ _____

14. $10 \times 9 =$ _____

15. $9 \times 2 =$ _____

16. $556 + 428 =$ _____

17. $849 - 158 =$ _____

Fluency Talk

How would you explain to a friend the patterns in the products of 5 and 10 that you can use to multiply fluently by 5 or 10?

How would you decompose by place value to subtract $456 - 278$?

Multiplication as Comparison

Focus Question

How can I compare using multiplication?

Hi, I'm Hannah.

I want to be a welder. I want to make a climber that is welded together—just like the one at the playground! I need 5 times as many pieces of pipe as I already have, so I need to understand multiplication as a comparison.

Name _____

Comparing Gardens

Compare the gardens and describe what you see.

Garden A

Garden B

Garden C

Garden D

Understand Comparing with Multiplication

Be Curious

What do you notice?
What do you wonder?

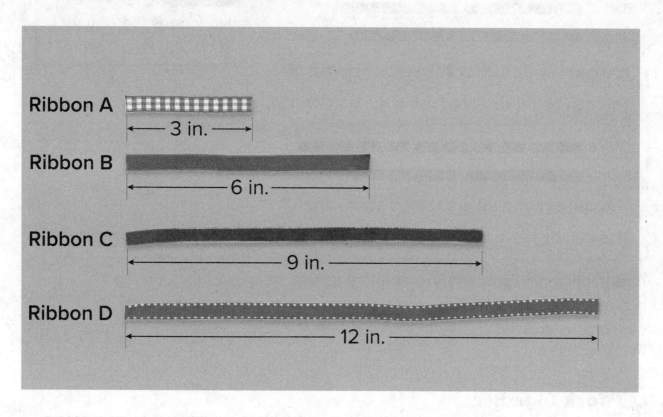

Ribbon A
← 3 in. →

Ribbon B
← 6 in. →

Ribbon C
← 9 in. →

Ribbon D
← 12 in. →

Math is... **Mindset**

How can your math skills or interests help you with your work today?

Learn

How can you describe the relationship between the number of cubes in Stick C and Stick D?

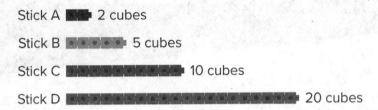

Stick A ▬▬ 2 cubes

Stick B ▬▬▬ 5 cubes

Stick C ▬▬▬▬▬ 10 cubes

Stick D ▬▬▬▬▬▬▬▬▬ 20 cubes

You can use multiplication to compare two quantities.

10 cubes 10 cubes Stick C ▬▬▬▬▬ ▬▬▬▬▬ Stick D ▬▬▬▬▬▬▬▬ 20 cubes	**Math is...** **Thinking** How does the equation show the relationship?

You can say that 20 is 2 *times as much as* 10.

This can be represented using the equation $20 = 2 \times 10$.

Stick A ▬ ▬ ▬ ▬ ▬ ▬ ▬ ▬ ▬ ▬

Stick D ▬▬▬▬▬▬▬▬ 20 cubes

You can say that 20 is 10 *times as much as* 2.

This can be represented using the equation $20 = 10 \times 2$.

A **multiplicative comparison** statement tells the relationship between two quantities using multiplication. You can use terms such as *times as many as* or *times as much as* to describe the relationship.

🔍 Work Together

Marvin collects pins from theme parks. He has 5 pins from the water park, 3 pins from the amusement park, and 15 pins from the wildlife park. What multiplicative comparison statements can you make about his pin collection? What equations can you write?

On My Own

Name _____

1. Which statement can be represented by the equation $3 \times 9 = 27$? Choose the correct answer.

   A. 3 is 3 times as much as 27.

   B. 27 is 9 times as much as 9.

   C. 3 is 9 times as much as 27.

   D. 27 is 3 times as much as 9.

2. Which statements are true? Choose all that apply.

   A. 9 is 2 times as much as 18.

   B. 2 is 9 times as much as 18.

   C. 18 is 2 times as much as 9.

   D. 9 is 18 times as much as 2.

   E. 18 is 9 times as much as 2.

3. Complete the multiplicative comparison statement.

   Stick A ▪▪▪▪▪▪▪▪▪▪▪▪▪

   Stick B ▪▪▪▪▪ ▪▪▪▪▪ ▪▪▪▪

   There are _____ times as many cubes in Stick A as in Stick B.

4. What multiplicative comparison statement can you write about the number of cubes in the two sticks?

   ▪▪ 2 cubes

   ▪▪▪▪▪▪▪▪▪▪ 10 cubes

5. What equation can be used to represent the multiplicative comparison statement 24 is 4 times as much as 6?

6. Sarah and Mark are looking at the equation $450 = 90 \times 5$. Sarah says it means 450 is 90 times as much as 5. Mark says it means 450 is 5 times as much as 90. How do you respond to them?

How can you draw pictures to represent each statement?

7. 16 is 4 times as many as 4.

8. 12 is 2 times as many as 6.

9. 12 is 3 times as many as 4 and 4 times as many as 3.

10. What equation can be used to represent 36 is 9 times as many as 4 and 4 times as many as 9?

11. **STEM Connection** Welding fuel comes in bottles of many sizes. The smallest bottle weighs 8 pounds, and the largest bottle weighs 9 times as much. What equation can you write to show how much the largest bottle weighs? Explain your answer.

12. **Extend Your Thinking** Aaron plants a garden with 4 tomato plants, 3 times as many pepper plants as tomato plants, and twice as many zucchini plants as pepper plants. Write equations to show how many pepper plants and zucchini plants are in the garden. How many plants does Aaron plant in all?

🎨 **Reflect**

What do you notice or think about when you use comparison words to describe a multiplication equation?

Math is... **Mindset**

How have your skills and interests helped you with your work today?

Represent Comparison Problems

? Be Curious

How are they the same?
How are they different?

Julie's shells

Mark has 4 times as many shells as Julie.

Jamal has 4 more shells than Julie.

> **Math is...** ❮ **Mindset**
> How can working as a team help you accomplish your goal?

Learn

Jackson biked 6 miles. Karen biked 4 more miles than Jackson. Terry biked 4 times as many miles as Jackson.

How many miles did Karen and Terry bike?

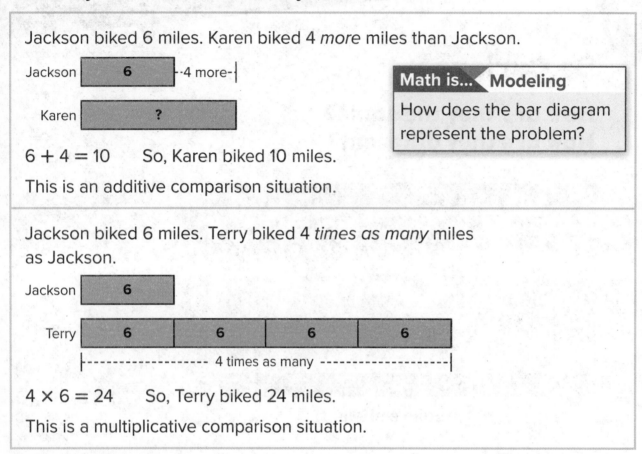

Jackson biked 6 miles. Karen biked 4 *more* miles than Jackson.

Jackson | 6 | --4 more--|

Karen | ? |

$6 + 4 = 10$ So, Karen biked 10 miles.

This is an additive comparison situation.

Math is... Modeling

How does the bar diagram represent the problem?

Jackson biked 6 miles. Terry biked 4 *times as many* miles as Jackson.

Jackson | 6 |

Terry | 6 | 6 | 6 | 6 |

|-------------- 4 times as many --------------|

$4 \times 6 = 24$ So, Terry biked 24 miles.

This is a multiplicative comparison situation.

Additive comparison statements compare quantities by telling how many more. They are represented by addition equations.

Multiplicative comparison statements compare quantities by telling how many times. They are represented by multiplication equations.

Work Together

Elliot scored 9 points in a basketball game. Michael scored 2 times as many points as Elliot, and Deanna scored 6 more points than Elliot. How many points did Michael and Deanna score? Use equations to represent the problem.

On My Own

Name _____

What equation can you write to represent and solve the comparison?

1. 8 more than 4

2. 3 times as many as 5

3. 2 times as long as 9 feet

4. 5 times as far as 10 miles

How can you represent the problem? Draw a bar diagram and write an equation to solve.

5. A small bridge is 40 feet long. A new bridge is 3 times as long as the small bridge. How long is the new bridge?

6. Raya has 8 pencils in her school box. Miranda has 4 more pencils than Raya. How many pencils does Miranda have?

7. Louisa is 5 feet tall. The tree in her backyard is 4 times as tall as Louisa. How tall is the tree?

8. Ameer planted 6 plants. David planted 5 times as many. How many plants did David plant? Write an equation to represent the problem.

9. Rosa and her brother are playing a game. Rosa scored 8 points and her brother scored 2 points. What are two comparison statements you can make about their scores?

10. A cat's tail can be 10 inches long. A lion's tail can be 3 times as long. How long can a lion's tail be? Write an equation to represent the problem.

11. Error Analysis Naomi wants to place three bookcases along a 7-foot wall. Each bookcase is 4 feet wide. She says the bookcases will fit, since 7 feet is three times longer than 4 feet. What would you tell her?

12. Extend Your Thinking Write a comparison problem for the equation $9 + 7 = ?$ and another for the equation $9 \times 7 = ?$. Then solve each problem.

⏱ **Reflect**

What differences did you notice between multiplicative and additive comparison problems as you solved them?

Math is... **Mindset**

How did working as a team help you accomplish your goal?

Be Curious

What math do you see?

Penelope and Madison are pitching at a softball tournament.
Penelope strikes out 3 times as many batters as Madison.
How many batters could Penelope have struck out?

Math is... Mindset

What steps can you take to focus on your work today?

Learn

Penelope strikes out 21 batters at the softball tournament. She strikes out 3 times as many batters as Madison.

How many batters does Madison strike out?

You can represent the problem with a bar diagram and an equation. Penelope strikes out 3 times as many batters as Madison.

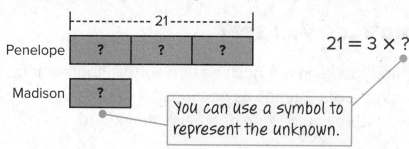

$21 = 3 \times ?$

You can use a symbol to represent the unknown.

Use multiplication to solve for the unknown.

$21 = 3 \times ?$

$21 = 3 \times 7$

Madison strikes out 7 batters.

Math is... **Modeling**

How do the bar diagram and equation represent the situation?

Multiplicative comparison problems can be solved using multiplication equations. These problems use words related to multiplication, such as *times as many* and *times as much*.

Work Together

Ty has 32 coins in his collection. Dan has 8 coins in his collection. How many times as many coins does Ty have than Dan? Use a bar diagram and an equation to represent and solve the problem.

On My Own

Name _____

What is the unknown number? Write a multiplication equation to represent the comparison. Then solve the equation.

1. 56 is ? times as much as 7.

2. 35 is 7 times as many as ?.

3. 24 is 8 times as many as ?.

4. 45 is ? times as much as 9.

How can you represent the problem? Draw a bar diagram and write a multiplication equation to solve.

5. Marie read 20 pages of a book last week. She read 2 times as many pages this week as she did last week. How many pages did she read this week?

6. A tomato plant is 48 inches tall. How many times as tall is the tomato plant as a pepper plant that is 8 inches tall?

7. Dana saved $63. Dana saved 7 times as much as Julie. How much did Julie save?

8. Wilani has 12 nickels. Wilani has 6 times as many nickels as Brenda. What is the value of all the coins Wilani and Brenda have? Explain your reasoning.

9. Perry ran 5 times as many minutes as Louis. How many minutes could Perry and Louis have run? Explain your answer.

10. **STEM Connection** A welder used 4 meters of metal rod last week and 32 meters of metal rod this week. How many times as many meters of metal rod did the welder use this week compared to last week? Write an equation to represent and solve the problem.

11. There are 12 birds in the apple tree. This is 4 times as many birds as there are in the cherry tree. How many birds are in the cherry tree? Show your work.

12. **Extend Your Thinking** Vela practiced piano 12 hours last week. Vela practiced 4 times as long as Marina. Rian practiced 2 times as long as Marina. How does the time Vela practiced compare to the time Rian practied? Explain your reasoning.

🕐 **Reflect**

What comparison words do you look for in a problem to determine the equation you can use?

Math is... **Mindset**

What steps did you take to focus on your work today?

Solve Comparison Problems Using Division

Be Curious

Tell me everything you can.

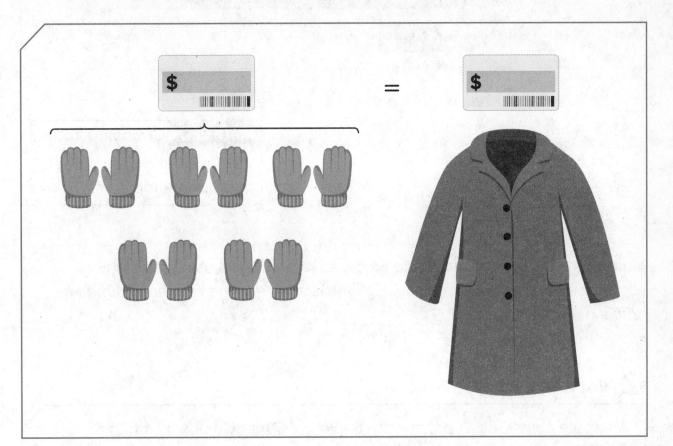

Math is... Mindset

How can creative thinking help you solve a problem?

Learn

A store sells winter coats for $50. A winter coat costs 5 times as much as a pair of gloves.

How much does a pair of gloves cost?

Use a bar diagram to represent the problem.
The winter coat costs 5 times as much as a pair of gloves.

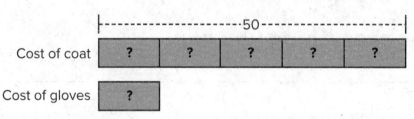

Write a division equation to represent and solve the problem.

$50 \div 5 = ?$

$50 = 5 \times 10$

$50 \div 5 = 10$

A pair of gloves costs $10.

Math is... Choosing Tools

Why is a bar diagram an appropriate tool?

Multiplicative comparison problems can also be solved using division equations. These problems may use words such as *times as many, times as much,* or *times less than*.

Work Together

An apple costs 36¢. A banana costs 12¢. How many times as much does an apple cost compared to a banana? Use a bar diagram and an equation to represent and solve the problem.

On My Own

Name _____

What is the unknown number? Write a division equation to represent the comparison. Then solve the equation.

1. 24 is 8 times as much as ?.

2. 20 is ? times as much as 5.

3. 18 is ? times as much as 6.

4. 16 is 4 times as much as ?.

How can you represent the problem? Draw a bar diagram and write a division equation to solve.

5. A piece of green string is 48 inches long. How many times as long is the green string than a piece of red string that is 8 inches long?

6. Ellie has 50 blue blocks. She has 5 times as many blue blocks as white blocks. How many white blocks does she have?

7. Charlie read 4 times as many pages as his sister. Charlie read 36 pages of his book. How many pages did Charlie's sister read? What equations represent the problem? Choose all that apply.

   A. $36 + 4 = ?$

   B. $36 - 4 = ?$

   C. $4 \times ? = 36$

   D. $4 \times 36 = ?$

   E. $? \div 4 = 36$

   F. $36 \div 4 = ?$

8. **Error Analysis** Michael scored 80 points on a video game. Alicia says his score is 4 times as much as her score. Michael thinks Alicia scored 320 points. How would you respond to him?

9. A rectangular garden is 3 times as long as it is wide. The length of the garden is 9 feet. How wide is the garden?

10. John ran 18 laps around the track. Sabrina ran 5 laps around the track. John ran twice as far as Mika and Sabrina combined. How many laps did Mika run around the track? Explain.

11. Cory learned that the airport is 5 times farther from his home than the library. He knows the airport is 30 miles from home. What is the distance from Cory's home to the library?

12. **Extend Your Thinking** Write a word problem about a multiplicative comparison that you can solve using the equation $35 \div 5 = ?$. Then solve.

Reflect

When you solve problems like the ones in this lesson, how do you know you need to divide?

Math is... Mindset

How has creative thinking helped you solve a problem?

Comparison Problems

Name _____

Read each word problem. Choose *all* equations that represent the problem. Do not actually solve the problem.

1. Ms. Olson is gathering materials to make supply boxes for her classroom. She has 9 pens. She also has 3 times as many pencils as pens. How many pencils does she have?

   Choose all that apply.

   a. $9 \times 3 = ?$

   b. $9 + 3 = ?$

   c. $9 \div 3 = ?$

   d. $9 + 9 + 9 = ?$

   Explain why you chose the equation or equations.

---

2. Mr. Gomez is gathering materials to make supply boxes for his classroom. He has 12 rulers. He also has 4 more erasers than rulers. How many erasers does he have?

   Choose all that apply.

   a. $12 \times 4 = ?$

   b. $12 + 4 = ?$

   c. $12 \div 4 = ?$

   d. $12 + 12 + 12 + 12 = ?$

   Explain why you chose the equation or equations.

3. The 4th-grade class is selling hats and shirts to raise money for a new fish tank. So far, they have sold 27 hats and 9 shirts. How many times as many hats as shirts have they sold?

Choose all that apply.

a. $27 \times 9 = ?$

b. $9 \times ? = 27$

c. $27 \div 9 = ?$

d. $9 + ? = 27$

Explain why you chose the equation or equations.

---

4. Raj looks for birds and dogs during his walk. Today he saw 24 birds. This is 6 times as many as the number of dogs he saw. How many dogs did Raj see?

Choose all that apply.

a. $24 \times 6 = ?$

b. $? \times 6 = 24$

c. $24 \div 6 = ?$

d. $6 + ? = 24$

Explain why you chose the equation or equations.

Reflect On Your Learning

I am confused.   I'm still learning.   I understand.   I can teach someone else.

Unit Review Name _____

Vocabulary Review

Choose the correct word(s) to complete the sentence.

additive comparison	multiplicative comparison

1. A(n) _____ statement compares two quantities using multiplication where one is times as much as, or times as many as the other. (Lesson 4-1)

2. "Sienna swam 2 more laps than Seth" is an example of a(n)

   _____. (Lesson 4-2)

3. A(n) _____ statement compares two quantities with addition where one is more or less than the other. (Lesson 4-2)

4. "Sienna swam 2 times as many laps as Seth" is an example of

   a(n) _____. (Lesson 4-2)

Review

5. Which comparison statement represents the equation $5 \times 3 = 15$? Choose the correct answer. (Lesson 4-1)

   A. 15 is 5 times as many as 3.

   B. 3 is 5 times as many as 15.

   C. 5 is 3 times as many as 15.

   D. 5 is 15 times as many as 3.

6. Tom and Gail are serving bowls of soup at a restaurant. Which comparison statement could be represented by the bar diagram? Choose the correct answer.
(Lesson 4-2)

Tom	3 bowls		
Gail	3 bowls	3 bowls	3 bowls

   A. Tom serves 3 times as many bowls of soup as Gail.

   B. Tom serves 3 more bowls of soup than Gail.

   C. Gail serves 3 more bowls of soup than Tom.

   D. Gail serves 3 times as many bowls of soup as Tom.

7. Maria saved 6 times as much money as Ali. How much money could Maria and Ali have saved? Justify your answer. (Lesson 4-3)

8. What expression represents the situation? Draw a line to match them. (Lessons 4-2, 4-3, 4-4)

Vigo eats 8 cherries. Dina eats 4 more than Vigo. How many cherries does Dina eat? $8 + 4$

Cody stacks 8 boxes. This is 4 times as many boxes as Audrey stacks. How many boxes does Audrey stack? $8 - 4$

There are 8 flowers growing in a pot. There are 4 times as many growing in the garden. How many flowers are growing in the garden? 8×4

 $8 \div 4$

9. Amahle has 30 pairs of shoes. That is 3 times as many as her 10 pairs of jeans. What equation represents this comparison? (Lesson 4-1)

10. Keisha, Bryan, and Hiran are playing basketball. Keisha made 6 shots. Bryan made 2 times as many shots as Keisha. Hiran made 3 more shots than Keisha. How many shots did Bryan and Hiran make? (Lesson 4-2)

11. Paula sends 5 times as many texts as her mother. Paula sends 40 texts. How many texts does her mother send? (Lesson 4-4)

   a. Which equation represents the problem? Choose the correct answer.

   A. $? + 5 = 40$

   B. $40 \times 5 = ?$

   C. $40 + 5 = ?$

   D. $40 \div 5 = ?$

   b. What is the solution to the problem?

12. What are two multiplicative comparisons that represent the equation $4 \times 2 = 8$? Complete each statement. (Lessons 4-1)

   _____ is _____ times as many

   as _____ .

   _____ is _____ times as many

   as _____ .

13. During a trip to the beach, Cheri collects 4 times as many seashells as Natalie. Cheri collects 24 seashells. How many seashells did Natalie collect? (Lessons 4-3, 4-4)

   a. What equation can you write to represent the problem? Use ? to represent the unknown.

   b. What is the solution to the problem?

14. Jeong-Li has 6 pennies, 8 dimes, and 24 nickels. (Lessons 4-3, 4-4)

   a. How many times as many nickels does Jeong-Li have as dimes?

   b. How many times as many nickels does Jeong-Li have as pennies?

15. Carlos has 10 dog treats. He buys a box that has 5 times as many dog treats. How many dog treats are in the box? Explain your answer. (Lessons 4-3, 4-4)

Performance Task

Hannah and her dad are making a sculpture using welding rods that are 12 inches long, 14 inches long, and 18 inches long. They use five 18-inch rods, twice as many 14-inch as 18-inch rods, and ten more 12-inch rods than 18-inch rods to build the sculpture.

Part A: How many welding rods did Hannah and her dad use to build the sculpture? Explain how you found your answer.

Part B: How many times as many 12-inch rods did Hannah and her dad use as 18-inch rods? Explain.

Part C: Hannah and her dad used 3 times as many welding rods to build this sculpture than their last sculpture. How many welding rods did Hannah and her dad use on the last sculpture? Show your work.

🕐 Reflect

How is using multiplication to compare different from using addition to compare?

Fluency Practice

Name _____

Fluency Strategy

You can use strategies, such as doubling and breaking apart, to multiply.

You can double 2s facts to multiply by 4. You can double 4s facts to multiply by 8.

$5 \times 4 = ?$
$5 \times 2 = 10$, and 4 is double 2. So, 5×4 is double 5×2.
$5 \times 4 = 20$
$6 \times 8 = ?$
$6 \times 4 = 24$ and 8 is double 4. So, 6×8 is double 6×4.
$6 \times 8 = 48$

You can break apart a factor to multiply by 8.

Since $8 = 5 + 3$, $\quad 6 \times 8 = 6 \times 5 + 6 \times 3$
$6 \times 8 = 30 + 18$
$6 \times 8 = 48$

1. What is 9×8?

Since $8 = 3 +$ _____, $9 \times 8 = 9 \times$ _____ $+$ _____ $\times 3$

$9 \times 8 =$ _____ $+$ _____

$9 \times 8 = 72$

Fluency Flash

Use the model to complete the multiplication fact.

2.

3.

$3 \times 2 =$ _____ $\qquad$ $4 \times 4 =$ _____

So, $3 \times$ _____ $=$ _____ $\qquad$ So, $4 \times$ _____ $=$ _____

Fluency Check

What is the product or difference?

4. $3 \times 8 =$ _____

5. $4 \times 9 =$ _____

6. $847 - 246 =$ _____

7. $7 \times 8 =$ _____

8. $9 \times 5 =$ _____

9. $543 - 121 =$ _____

10. $4 \times 4 =$ _____

11. $7 \times 4 =$ _____

12. $359 - 157 =$ _____

13. $6 \times 2 =$ _____

14. $4 \times 6 =$ _____

15. $698 - 482 =$ _____

16. $8 \times 8 =$ _____

17. $10 \times 6 =$ _____

Fluency Talk

How can you explain to a friend how to use breaking apart to find an 8s multiplication fact?

How is multiplying by 4 related to multiplying by 2?

Numbers and Number Patterns

April

Focus Question

How can I use patterns to describe the
relationship between numbers?

Hi, I'm Haley.

I want to be an astronomer. I love observing the
phases of the moon. Do you know that there is a
pattern in the phases of the moon? How can we
determine that pattern?

STEM
video | GO
ONLINE

Name _____

What's in a Spiral?

Look at the picture. What do you notice?

Part A

Part B

Find the pattern. Then fill in the next numbers.

Hemachandra/Fibonacci sequence: 0, 1, 1, 2, 3, 5,

_____ , _____ , _____ , _____

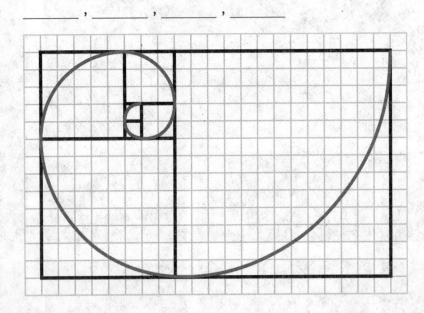

Understand Factors of a Number

Be Curious

What do you notice?
What do you wonder?

Math is... Mindset

What are your strengths in math?

Learn

Rachel has 18 stickers to arrange in rows on a poster. She wants to put the same number of stickers in each row.
How can she arrange the stickers?

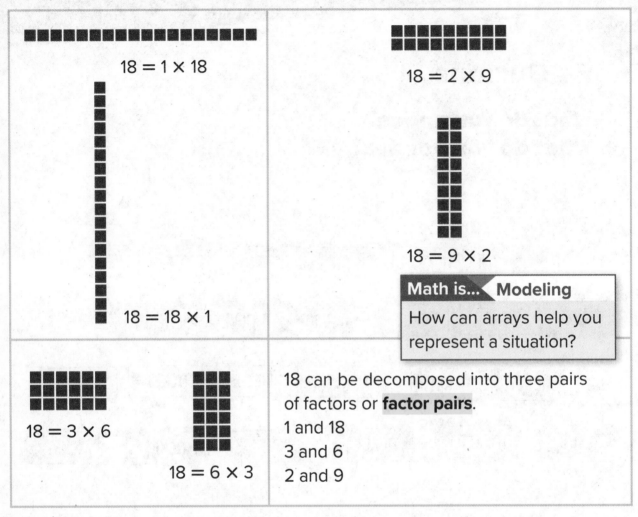

$18 = 1 \times 18$

$18 = 18 \times 1$

$18 = 2 \times 9$

$18 = 9 \times 2$

Math is... Modeling

How can arrays help you represent a situation?

$18 = 3 \times 6$

$18 = 6 \times 3$

18 can be decomposed into three pairs of factors or **factor pairs**.
1 and 18
3 and 6
2 and 9

All numbers have at least two factors called a factor pair. To find all the factor pairs of a number, start with 1 and then check each number to determine whether it is a factor, until the reverse pair is reached.

Work Together

What are the factor pairs of 48?

On My Own

Name _____

What are all the factor pairs for each number?

1. 14

2. 65

3. 23

4. 64

5. 32

6. 100

7. Adrian arranges 12 flowers. He puts the same number of flowers in each vase and can use up to 6 vases. What are two other ways to arrange the flowers?

8. Setsuko is organizing 36 books in her bookcase. She wants the same number of books on each shelf and can use up to 3 shelves. What are three different ways she can arrange her books?

9. The soccer coach has 24 trophies to display in a cabinet. How can she display the trophies in equal rows? Find all possible arrangements.

10. **STEM Connection** Finn used 84 nails to build
    7 shelves. Can Finn use 12 nails for each shelf?
    Explain your reasoning.

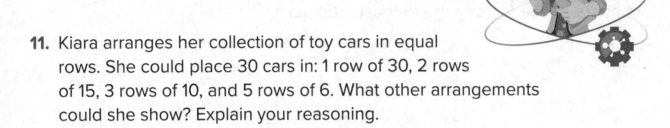

11. Kiara arranges her collection of toy cars in equal
    rows. She could place 30 cars in: 1 row of 30, 2 rows
    of 15, 3 rows of 10, and 5 rows of 6. What other arrangements
    could she show? Explain your reasoning.

12. Ahmet is planting flower bulbs in his garden. Can he arrange
    11 bulbs in 3 equal rows? Explain.

13. **Extend Your Thinking** What number less than 20 has 3 factor
    pairs, but only 5 factors? Explain.

Reflect

What strategy can you use to find all factor pairs of a number?

Math is... **Mindset**

How have you used your
strengths today?

Factors

Name _____

For each number provided, circle its factors. All the factors of the number may not be included.

1. Circle the numbers that are factors of **14**:

1	4	10	28
2	7	14	42

Explain your thinking.

2. Circle the numbers that are factors of **24**:

1	6	12	24
4	8	20	48

Explain your thinking.

For each number provided, circle its factors. All of a number's factors may not be included.

3. Circle the numbers that are factors of **27**:

1	3	8	27
2	7	14	42

Explain your thinking.

4. Circle the numbers that are factors of **19**:

1	6	10	19
3	9	13	38

Explain your thinking.

Reflect On Your Learning

I'm confused.   I'm still learning.   I understand.   I can teach someone else.

Understand Prime and Composite Numbers

? Be Curious

What do you notice?
What do you wonder?

2 balloons

5 balloons

8 balloons

9 balloons

> **Math is...** ◄ **Mindset**
>
> What are some ways you can contribute to your group today?

Learn

A sports store arranges some balls in equal rows on shelves.

How is the way they can arrange the basketballs different from the way they can arrange the soccer balls?

In stock: 12   In stock: 17

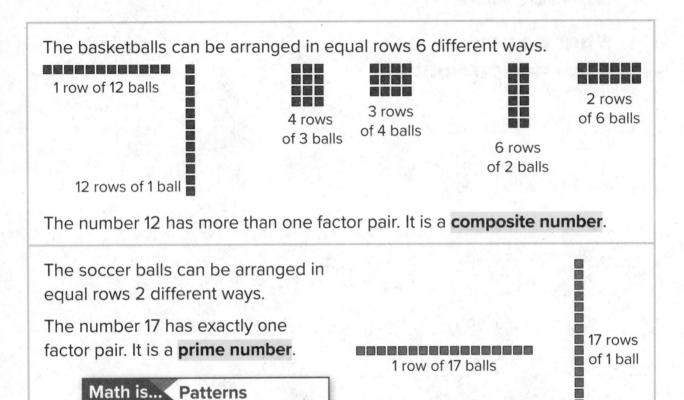

The basketballs can be arranged in equal rows 6 different ways.

1 row of 12 balls

12 rows of 1 ball

4 rows of 3 balls

3 rows of 4 balls

6 rows of 2 balls

2 rows of 6 balls

The number 12 has more than one factor pair. It is a **composite number**.

The soccer balls can be arranged in equal rows 2 different ways.

The number 17 has exactly one factor pair. It is a **prime number**.

1 row of 17 balls

17 rows of 1 ball

Math is... Patterns

What patterns do you notice in prime numbers?

You can classify a whole number as prime or composite based on the number of factor pairs it has.

Work Together

Which of these numbers are prime? Which numbers are composite?

8   19   33   45   67

On My Own

Name _____

Is the number prime or composite? Explain your reasoning.

1. 3

2. 24

3. 15

4. 31

5. 87

6. 2

Is the statement true or false? Justify your answer.

7. All even numbers greater than 2 are composite.

8. 1 is a prime number.

9. All odd numbers are prime.

10. All prime numbers are odd.

11. Find a prime number greater than 50. Explain how you know it is prime.

12. Find a composite number greater than 75. Explain how you know it is composite.

13. **Error Analysis** Scott says he can arrange 71 marbles into equal groups in more than 2 ways. Do you agree with Scott? Explain your reasoning.

14. **Extend Your Thinking** Armando's button collection is shown. He wants to arrange each type of button into equal groups. Which buttons will he be able to arrange into groups of 2 or more? Explain your reasoning.

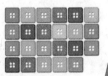

 **Reflect**

How are factor pairs related to prime and composite numbers?

Math is... Mindset

How did you contribute to your group today?

Be Curious

Which doesn't belong?

10

5

23

15

Math is... Mindset

How can you show others
you respect their ideas?

Learn

Ralph buys packs of water bottles for a sports event.

How many water bottles might he buy?

You can use multiples of 8 to determine the number of bottles Ralph might buy.

8, 16, 24, 32, and 40 are products of 8 and another whole number.

8, 16, 24, 32, and 40 are multiples of 8.

$1 \times 8 = 8$

$2 \times 8 = 16$

$3 \times 8 = 24$

$4 \times 8 = 32$

$5 \times 8 = 40$

Math is... Generalizations

Why can you say any whole number is a multiple of its factors?

1, 2, 4, and 8 are factors of 8.

8 is a multiple of each of its factors.

1: 1, 2, 3, 4, 5, 6, 7, ⑧
2: 2, 4, 6, ⑧
4: 4, ⑧
8: ⑧

A multiple of a whole number is the product of that number and another whole number. Any whole number is a multiple of each of its factors.

🗨 Work Together

Which of these numbers are factors of 70? Explain how you know.

2   3   4   5   10   12   13

On My Own

Name _____

What are the next five multiples of the number?

1. 4, _____, _____, _____, _____, _____

2. 7, _____, _____, _____, _____, _____

3. 12, _____, _____, _____, _____, _____

4. 15, _____, _____, _____, _____, _____

Choose all that apply.

5. Which numbers are multiples of 4?

   A. 14

   B. 16

   C. 34

   D. 64

6. Which numbers are multiples of 9?

   A. 91

   B. 89

   C. 45

   D. 18

What are the missing multiples?

7. _____, 12, 18, _____, _____, _____

8. _____, 10, _____, _____, _____

9. What do you know about the patterns in the products of 5? How can this help you determine if a number is a multiple of 5?

10. There are 3 olives on each slice of pizza. Danny will eat 2 or 3 slices. How many olives might Danny eat? Justify your thinking.

11. Extend Your Thinking Complete the Venn diagram by using numbers between 1 and 72. How can you describe the numbers shown in the overlapping section of the diagram?

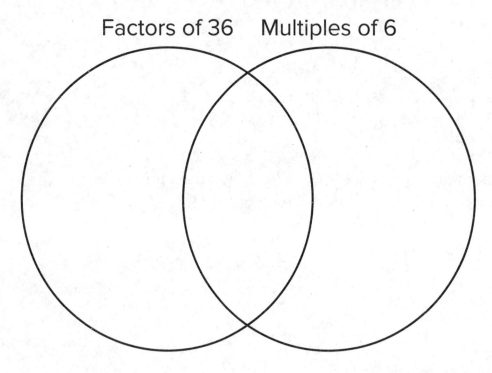

Factors of 36   Multiples of 6

 **Reflect**

How can you use patterns to decide whether a number is a multiple of another number?

Math is... **Mindset**

How have you shown others you respect their ideas?

Number or Shape Patterns

Be Curious

What do you notice?
What do you wonder?

Math is... Mindset

How can being flexible in your thinking help you make good decisions?

Learn

Opa is making necklaces.

If Opa continues the patterns, what will she place next on each necklace?

Necklace A

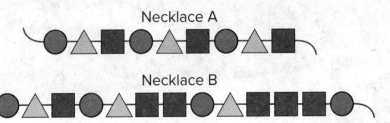

Necklace B

This **sequence** of shapes follows a repeating pattern.

The **pattern unit** is blue circle, yellow triangle, red square.

Opa will add a blue circle, a yellow triangle, and a red square to extend the pattern on Necklace A.

> **Math is...** **Patterns**
>
> What does it mean to say there is a pattern?

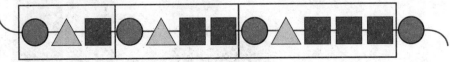

This sequence of shapes follows a growing pattern.

The **pattern rule** is add 1 more red square with each pattern unit.

Opa will add a blue circle, a yellow triangle, and four red squares to extend the pattern on Necklace B.

A repeating pattern has a pattern unit that defines the pattern.
A growing pattern has a rule that defines the pattern.

💬 Work Together

What are the next three numbers in the pattern? Explain how you know.

10, 14, 18, 22, ...

On My Own

Name _____

What is the pattern unit or rule?

1. ■■□○▲▲■■○▲

2. 6, 12, 24, 48, 96

3. 4, 8, 10, 4, 8, 10

4.

5.

6. 12, 20, 28, 36, 44

Extend the pattern to determine three more numbers or shapes in the pattern. How did you find your answer?

7. 36, 30, 24, _____, _____, _____

8.

9. _____, 4, 8, _____, 32, _____, 128

10.

11. Extend Your Thinking What would be the shape of the 30th bead in the string? Explain your thinking.

Reflect

How can you determine whether a sequence of shapes or numbers follows a pattern?

Generate a Pattern

Be Curious

Which doesn't belong?

1, 2, 4, 8, 16, ...

14, 12, 10, 8, 6, ...

7, 9, 11, 13, 15, ...

0, 4, 8, 12, 16, ...

Math is...   **Mindset**

What strategies help you
work more efficiently?

Learn

Thomas is on page 12 in his book. He plans to read 5 pages of his book each day this week.

What page will he be on after 5 days?

You can use a table to detemine the solution.

You can add groups of 5 pages to the 12 pages Thomas already read to find each number, or **term**, in the pattern.

Day	Rule	Pages Read
1	1 × 5 + 12	17
2	2 × 5 + 12	22
3	3 × 5 + 12	27
4	4 × 5 + 12	32
5	5 × 5 + 12	37

The 5th term in the pattern is 37.

Thomas will be on page 37 after 5 days.

Math is... Structure

How do each of the terms in the pattern relate to the one before it?

You can use a rule to create a pattern. The rule tells you how the pattern works.

Work Together

Jasmine makes a row of 2 counters. She makes another row using 1 more counter than in the previous row. If she continues this pattern, how many counters will there be in the 6th row?

On My Own

Name _____

What are the first five terms of the pattern? Write or draw them.

1. Starting with 40, subtract 4.

2. 2 circles, 1 square repeating

3. Starting with 14 dots in a row, subtract 2 dots from each row.

4. Starting with 3, add 6.

5. Hector puts 3 photos on the first page of his scrapbook. He increases the number of photos on each page by 3. How many photos are on the pages indicated?

Page	Rule	Number of photos
2	(1 × 3) + 3	6
4		
6		

6. An amphitheater has 5 seats in the floor rows. Each stair row has two more seats on each end than the previous row. How many seats are in the stair rows indicated?

Row	Rule	Number of seats
6		
8	(8 × 4) + 5	37
10		

7. Carton number 1 has 3 pairs of socks. Each carton has double the amount as the previous carton. How many socks are in each of the cartons?

Carton	Rule	Number of socks
1	1 × 3	
2		
3	4 × 3	
4		

8. Create a pattern with 5 terms. Describe the rule your pattern follows.

9. Frank is planting a tulip plant in each planter, alternating red, yellow, and white tulips. What color tulip is in each planter?

10. Lola travels 4 miles to school and back each day. How many miles does she travel in 5 days? In 20 days? Explain your reasoning.

11. Error Analysis Dante drew a pattern to represent the rule, *add 4 squares to each column, starting with 2*. His pattern had 18 squares in the 5th column. How can you explain whether or not his pattern follows the rule?

12. Extend Your Thinking A pattern has the rule, *add 6 starting with 8*. How much greater is the 6th term than the 4th term? Explain your reasoning.

⟳ Reflect

How can a pattern rule tell you what the pattern will look like?

Math is... **Mindset**

What strategies helped you work more efficiently?

Analyze Features of a Pattern

Be Curious

How are they the same?
How are they different?

| 24 | 30 | 36 | 42 | 48 |

| 13 | 19 | 25 | 31 | 37 |

Math is... Mindset

How can you know that you have made good decisions?

Learn

Charlie used craft sticks to create these triangles.

If he continues the pattern, will he use exactly 34 craft sticks to create a triangle?

You can analyze the pattern to solve the problem.

Use a table to extend and analyze the pattern.

Math is... Choosing Tools

Why is a table a useful tool to determine patterns?

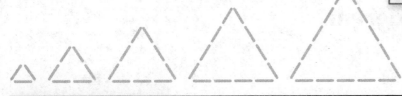

Triangle Number	1st	2nd	3rd	4th	5th
Number of Sticks	3	6	9	12	15

▶ **One Feature**

Each triangle has 3 more sticks than the previous triangle.

▶ **Another Feature**

The number of sticks in each triangle is a multiple of 3.

34 is not a multiple of 3, so Charlie will never use exactly 34 craft sticks.

You can analyze a pattern to find features that are not stated in the pattern rule.

🗨 Work Together

The flags on a banner follow a pattern. How can you use features of the pattern to determine the design of the 47th flag?

1st 2nd 3rd 4th 5th 6th 7th 8th

On My Own

Name _____

Is each statement about the pattern true or false?

1. Start with 4, multiply by 2.

	True	False
All terms are multiples of 4.		
All terms are multiples of 2.		
All terms are even numbers.		

2.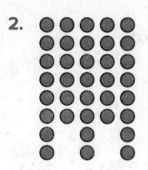

	True	False
The 16th column will have 8 dots.		
The number of dots in all terms is a multiple of 2.		
The number of dots increases in each term.		

3.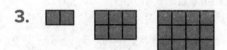

	True	False
All side lengths are multiples of 2.		
Each term increases by 1 row.		
The perimeter of each term is a multiple of 2.		

Use the pattern to answer exercises 4–6.

18, 24, 30, 36, 42, ...

4. What is the pattern rule?

5. What is a feature of the pattern that is not stated in the pattern rule? Explain why this feature exists.

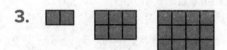

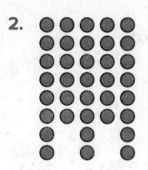

6. Can the number 72 be part of the pattern? Explain your answer.

18, 24, 30, 36, 42, ...

7. STEM Connection Haley is in charge of the telescope posts for students to view the stars. She sets up the telescopes in arrays at each post. Analyze her arrays to answer the questions.

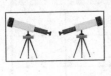  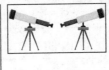 

a. What is the pattern rule Haley used to set up the telescopes?

b. What is a feature of the pattern not stated in the pattern rule?

c. If the pattern continues, explain what can you predict about the 18th post in the telescope arrangements?

8. Extend Your Thinking Stacy creates a square with 2-inch side lengths. She increases the length of the sides of the square by 1 inch as she makes each new square. If she continues this pattern, what would be the area of the 9th square? Explain.

🖐️ **Reflect**

How can you analyze a pattern to identify the features?

Math is... Mindset

How do you know that you made good decisions?

Unit Review
Name _____

Vocabulary Review

Choose the correct word(s) to complete the sentence.

composite number	factor pairs	pattern rule
prime number	multiple	sequence
term		

1. 3, 6, 9, 12, 15, ... is an example of a(n) _____
   (Lesson 5-4)

2. A(n) _____ is a whole number that has more
   than one factor pair. (Lesson 5-2)

3. A(n) _____ describes a pattern and can be
   used to extend a sequence of numbers or shapes. (Lesson 5-4)

4. A(n) _____ is a set of 2 factors that are
   multiplied to make a given product. (Lesson 5-1)

5. A(n) _____ is a whole number with exactly
   one factor pair. (Lesson 5-2)

6. A(n) _____ is a number in a sequence.
   (Lesson 5-5)

7. A(n) _____ of a whole number is a product of
   the number and another whole number. (Lesson 5-3)

Review

8. Alvin had 4 picture frames to arrange in rows on his wall. He wants to put the same number of picture frames in each row. Show how he can arrange the picture frames. (Lesson 5-1)

9. Use equations to represent different ways to arrange 6 water bottles. (Lesson 5-1)

10. List the factor pairs of 88. (Lesson 5-1)

11. A number has the factor pair 4 and 8. What is the number? Is the number prime or composite? (Lesson 5-2)

_____

12. How can you determine if the number 11 is prime or composite. Explain your reasoning. (Lesson 5-2)

13. Which numbers are multiples of 6? Choose all that apply. (Lesson 5-3)

   A. 16

   B. 30

   C. 34

   D. 36

   E. 42

14. Which numbers have 24 as a multiple? Choose all that apply. (Lesson 5-3)

   A. 3

   B. 4

   C. 9

   D. 12

   E. 20

15. Identify the pattern rule for the sequence of numbers. (Lesson 5-4)

   7   14   28   56   112

   _____

16. Identify the pattern rule and extend the sequence. (Lesson 5-4)

11, 23, 35, 47, _____, _____, _____

Pattern Rule: _____

17. Draw the first ten terms for the pattern from the given pattern unit. 3 stars, 1 heart (Lesson 5-5)

18. During the first week, Frida saves 20 dollars. Each week after, she saves 8 dollars. How much will she have by the fifth week? Eighth week? Tenth week? (Lesson 5-5)

19. A numeric pattern starts with the number 2 and follows the rule "add 4". Identify a feature of this pattern. Then use the feature to explain whether the number 79 will be a term in the pattern. (Lesson 5-6)

20. Which statements are true about the pattern? Choose all that apply. (Lesson 5-6)

A. The pattern grows by an even number of dots.

B. The pattern grows by an odd number of dots.

C. Each term has 2 more dots in a row.

D. Each term has 1 more row of dots than the previous term.

E. Each term has 2 more rows of dots than the previous term.

21. Mrs. Cooper is setting up her classroom. She has 21 students. Can she arrange the chairs in equal groups of 4? Explain. (Lesson 5-1)

Performance Task

Hayley wants to spend the same amount of time studying each star, study more than 1 star, and spend more than 1 hour studying each.

Part A: In January, Hayley has 24 hours to study stars. How many stars might she study? Explain.

Part B: Each month, Haley increases the number of hours she has for studying stars by 30 hours. How much time does Haley have to study stars each month for the next five months? Explain.

Part C: What is the longest time she might study a star during the first half of the year? Explain.

Reflect

What strategies can be used to analyze number and shape patterns?

Unit 5
Fluency Practice

Name _____

Fluency Strategy

You can use strategies such as breaking apart and doubling to multiply by 3 and 6.

Break apart 3.	Doubling.	Break apart 6.
$3 = 2 + 1$	$5 \times 6 = ?$	$6 = 5 + 1$
$5 \times 3 = 5 \times 2 + 5 \times 1$	$5 \times 3 = 15$	$5 \times 6 = 5 \times 5 + 5 \times 1$
$5 \times 3 = 10 + 5$	6 is double 3.	$5 \times 6 = 25 + 5$
$5 \times 3 = 15$	So, $5 \times 6 = 30$	$5 \times 6 = 30$

1. Find 7×3.

Since $3 = 2 +$ _____,

$7 \times 3 = 7 \times$ _____ + _____ $\times 1$

$7 \times 3 =$ _____ + _____

$7 \times 3 =$

Fluency Flash

Use the models to complete the multiplication facts.

2.

$4 \times 3 = 4 \times$ _____ + _____ $\times 1$

$4 \times 3 =$ _____ + _____

$4 \times 3 =$ _____

3.

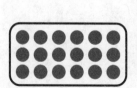 

$3 \times 6 =$ _____

So, $6 \times 6 =$ _____

Fluency Check

Find the product.

4. $7 \times 6 =$ _____

5. $3 \times 9 =$ _____

6. $2 \times 7 =$ _____

7. $12 \times 4 =$ _____

8. $8 \times 3 =$ _____

9. $9 \times 6 =$ _____

10. $2 \times 8 =$ _____

11. $4 \times 9 =$ _____

12. $6 \times 7 =$ _____

13. $5 \times 8 =$ _____

14. $8 \times 7 =$ _____

15. $6 \times 8 =$ _____

16. $10 \times 9 =$ _____

17. $4 \times 3 =$ _____

Fluency Talk

Explain to a friend how to multiply 3×8 by breaking apart 3.

Explain how multiplying by 6 is similar to multiplying by 8.

Multiplication Strategies with Multi-Digit Numbers

Focus Question

How can I multiply multi-digit numbers?

Hi, I'm Maya.

I want to be a geologist. I found a rock that is five thousand years old—and a rock that is one hundred times as old! I can multiply multi-digit numbers to find the age of the second rock. Multiplication makes my job easier!

Name _____

Area Puzzles

Find the missing areas and record them in the spaces.

All lengths and widths of the rectangles are whole numbers.

Puzzle A

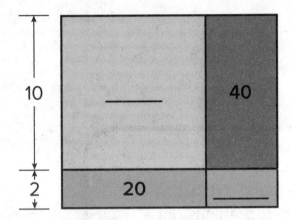

Puzzle B

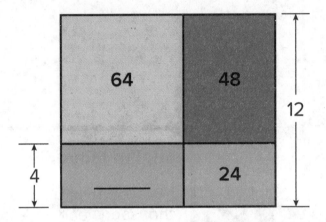

Puzzle C

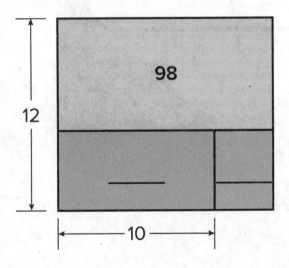

Puzzle D

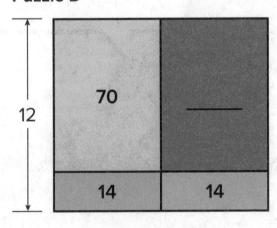

Multiply by Multiples of 10, 100, or 1,000

Be Curious

How are they the same?
How are they different?

about 2 pounds

about 20 pounds

about 200 pounds

about 2,000 pounds

Math is... Mindset

What do you do to contribute to the classroom community?

Learn

A bakery produces 4 types of bread every day. They make 300 loaves of each type.

How many loaves of bread do they make each day?

You can use a multiplication equation to represent and solve the problem.

▶ **One Way** Use basic facts and place value.

4 groups of 300

$n = 4 \times 3$ hundreds

$4 \times 3 = 12$

4×3 hundreds $= 12$ hundreds

$n = 1,200$

▶ **Another Way** Use the **Associative Property of Multiplication.**

Decompose 300.

$n = 4 \times 3 \times 100$

$n = 12 \times 100$

$n = 1,200$

Math is... Structure

How can you use patterns to find products?

$4 \times 30 = 12$ tens
$4 \times 30 = 120$

$4 \times 300 = 12$ hundreds
$4 \times 300 = 1,200$

$4 \times 3,000 = 12$ thousands
$4 \times 3,000 = 12,000$

When multiplying a whole number by a multiple of 10, 100, or 1,000, you can use place value to determine the number of zeros in the product.

Work Together

How can you solve the equation?

$8 \times 2,000 = ?$

On My Own

Name _____

What's the product? Complete the equation.

1. $4 \times 40 = 4 \times$ _____ tens

   $=$ _____ tens

   $=$ _____

2. $4 \times 400 = 4 \times$ _____ hundreds

   $=$ _____ hundreds

   $=$ _____

3. $6 \times 600 = 6 \times$ _____

   $= 36$ _____

   $=$ _____

4. $6 \times 6,000 = 6 \times$ _____

   $= 36$ _____

   $=$ _____

5. $4 \times 20 = 4 \times 2 \times$ _____

   $=$ _____ $\times$ _____

   $=$ _____

6. $4 \times 200 = 4 \times 2 \times$ _____

   $=$ _____ $\times$ _____

   $=$ _____

7. $7 \times 300 =$ _____

8. $2 \times 900 =$ _____

9. $8 \times 80 =$ _____

10. $9 \times 7,000 =$ _____

11. **STEM Connection** Maya has 6 boxes of
glass beakers. There are 50 beakers in each box.
How many beakers are there? Show your work.

12. A company bought 4 trucks. The weight of each truck is
7,000 pounds. What is the total weight of the trucks?
Explain your work.

13. **Extend Your Thinking** A restaurant owner orders 3 boxes of
salt packets. There are 20,000 packets in each box. How many
packets does the restaurant owner order? Show your work.

Reflect

What patterns do you see when you multiply with multiples of 10, 100,
and 1,000? Explain.

Math is... Mindset

How did you help build the
classroom community today?

Be Curious

What could the question be?

Leon said there are about 5,400 minutes until the eclipse.

FIRST ALERT: TOTAL LUNAR ECLIPSE

87 hours until total eclipse!

Math is... Mindset

What actions can help you achieve your day's goal?

Learn

An apartment building has 262 apartments. There are 3 sinks in each apartment.

About how many sinks are there in the building?

You can use different strategies to estimate a product.

▶ **One Way** Use compatible numbers.

Choose a compatible number close to 262.

$$262 \times 3 = s$$
$$\downarrow$$
$$250 \times 3 = 750$$

There are about 750 sinks in the building.

▶ **Another Way** Use rounding.

Round 262 to the nearest hundred.

$$262 \times 3 = s$$
$$\downarrow$$
$$300 \times 3 = 900$$

There are about 900 sinks in the building.

Math is... Thinking

How can you decide which estimation strategy to use?

Some problems do not require an exact solution. You can find an estimated solution using different strategies.

💬 Work Together

What is an estimated solution? Use two different estimation strategies.

$$168 \times 5 = b$$

On My Own

Name _____

How can you use compatible numbers to estimate the product? Complete the equation.

1. $323 \times 5 = ?$

Estimated product:

_____ $\times 5 =$ _____

2. $146 \times 3 = ?$

Estimated product:

_____ $\times 3 =$ _____

3. $436 \times 5 = ?$

Estimated product:

_____ $\times 5 =$ _____

4. $6 \times 1{,}252 = ?$

Estimated product:

$6 \times$ _____ $=$ _____

How can you use rounding to estimate each product? Complete the equation.

5. $247 \times 7 = ?$

Estimated product:

_____ $\times 7 =$ _____

6. $396 \times 8 = ?$

Estimated product:

_____ $\times 8 =$ _____

7. $5 \times 448 = ?$

Estimated product:

$5 \times$ _____ $=$ _____

8. $3{,}456 \times 2 = ?$

Estimated product:

_____ $\times 2 =$ _____

How can you find the estimated product? Write an equation to show your work.

9. A school cafeteria serves 2,750 lunches each week. About how many lunches are served in 4 weeks?

10. Penny's Pencils produces 5,980 pencils each day. About how many pencils does the company produce in 5 days.

11. The school store has some boxes containing school supplies.

	Number of boxes	Number of items in a box
Notebooks	9	28
Scissors	8	275
Pencils	6	3,830

   a. About how many notebooks are there?

   b. About how many scissors are there?

   c. About how many pencils are there?

12. **Extend Your Thinking** What are some numbers you could multiply by 5 to get an estimated product of 1,500? List 3 different numbers.

Reflect

How do you think like a mathematician when you estimate products?

Math is... **Mindset**

How did your actions help you achieve your day's goal?

Use the Distributive Property to Multiply

Be Curious

Is it always true?

$$5 \times 6 = 5 \times 4 + 5 \times 2$$

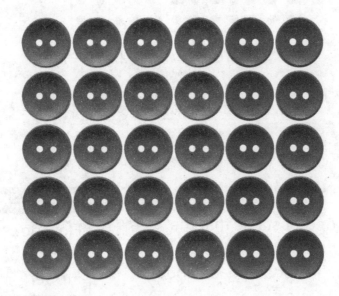

Math is... Mindset

How can you recognize and respond to the emotions of others?

Learn

What are some ways to solve the equation?

7 × 12 = ?

You can use the Distributive Property to solve multiplication equations.

When you use the Distributive Property, you decompose one factor. Then, you determine the product of each section, or the **partial products.**

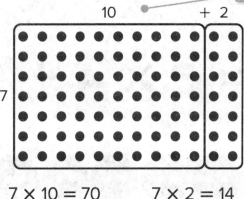

Decompose 12 to 10 + 2.

10   + 2

7

7 × 10 = 70   7 × 2 = 14

Each is a part of the product.

Add the two partial products to get the total product.

70 + 14 = 84

The Distributive Property states that the product of two factors is equal to the sum of the products of one factor and each addend of the decomposed factor.

Math is... **Structure**

How does an array help show the Distributive Property?

💬 Work Together

How can Leanne find the product of 6 × 15? Use the Distributive Property to show two different ways Leanne can find the product.

On My Own

Name _____

How can you use the Distributive Property to find the product?
Use the array to help you decompose and complete the equation.

1. $5 \times 13 = 5 \times (\underline{} + \underline{})$

$= (5 \times \underline{}) + (5 \times \underline{})$

$= \underline{} + \underline{}$

$= \underline{}$

2. $4 \times 15 = 4 \times (\underline{} + \underline{})$

$= (4 \times \underline{}) + (4 \times \underline{})$

$= \underline{} + \underline{}$

$= \underline{}$

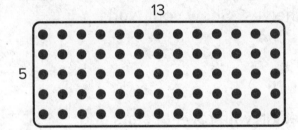

How can you use the Distributive Property to find the product?
Write and solve an equation to show your work.

3. 7×9

4. 12×8

5. 3×14

6. 5×17

7. **Error Analysis** Quin says he can find 6 × 8 by using (3 × 8) + (3 × 8). Jasmine says she can find 6 × 8 by using (6 × 4) + (6 × 4). Who is correct? Explain.

8. Kayla planted 6 rows of flower bulbs. There are 13 bulbs in each row. How many bulbs did she plant? Show your work.

9. A stock room has 4 shelves. Each shelf can hold 14 boxes. How many boxes can be stored on the shelves? Show your work.

10. **Extend Your Thinking** A pillow has a design with rows of stars and squares. There are 7 stars and 8 squares in each row. The equation 35 + 40 = 75 represents the number of stars and squares. How can you use the equation and the Distributive Property to find the number of rows on the pillow? Show your work.

Reflect

How can you find products using the Distributive Property?

Math is... Mindset

How have you worked to recognize and respond to the emotions of others?

Multiply 2-Digit by 1-Digit Factors

? Be Curious

What math do you see?

Math is... Mindset

What helps you focus when you feel frustrated?

Learn

What is the area of Zoe's lawn?

You can use the Distributive Property to solve the problem.

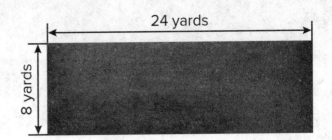

8 yards

24 yards

You can draw an area model.

An **area model** is a rectangle partitioned into sections.

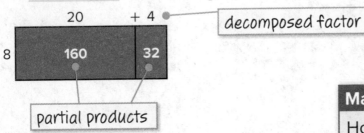

20   + 4 •———— decomposed factor

8   160   32

partial products

Add the partial products.

$160 + 32 = 192$

The area of Zoe's lawn is 192 square yards.

> **Math is...** **Choosing Tools**
>
> How do area models represent the decomposed factors?

To find a product of a 2-digit factor and a 1-digit factor, you can decompose the 2-digit factor using place value. Then find the partial products and add them.

🗨 Work Together

Toby is painting a fence that is 6 feet high and 52 feet long. What is the area of the fence? Solve using an area model and partial products.

On My Own

Name _____

How can you decompose the 2-digit number to solve? Complete the equation.

1. $6 \times 82 = ?$
$6 \times ($ _____ $+$ _____ $) = ?$

2. $91 \times 8 = ?$
$8 \times ($ _____ $+$ _____ $) = ?$

3. $76 \times 3 = ?$
$3 \times ($ _____ $+$ _____ $) = ?$

4. $7 \times 45 = ?$
$7 \times ($ _____ $+$ _____ $) = ?$

How can you decompose a factor and find the partial products? Complete the area model and equation to show your work.

5. 7×52

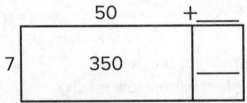

$7 \times 52 = (7 \times 50) + (7 \times$ _____ $)$
$7 \times 52 = 350 +$ _____
$7 \times 52 =$ _____

6. 4×96

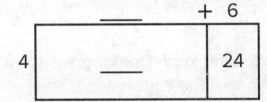

$4 \times 96 = (4 \times$ _____ $) + (4 \times 6)$
$4 \times 96 =$ _____ $+ 24$
$4 \times 96 =$ _____

7. 5×47

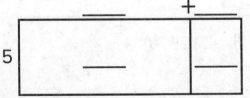

$5 \times 47 = (5 \times$ _____ $) + (5 \times$ _____ $)$
$5 \times 47 =$ _____ $+$ _____
$5 \times 47 =$ _____

8. 3×29

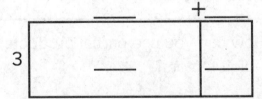

$3 \times 29 = (3 \times$ _____ $) + (3 \times$ _____ $)$
$3 \times 29 =$ _____ $+$ _____
$3 \times 29 =$ _____

9. **STEM Connection** Maya is looking for crystals in a dry streambed. She has collected samples from a region that measures 8 meters wide and 84 meters long. What is the area of the streambed she has searched?

10. Greg is getting stone blocks for a building project. There are 6 blocks, each weighing 53 kilograms. What is the total weight of the blocks?

11. **Extend Your Thinking** Marco made a curtain for the school stage. The curtain was 9 feet high and 112 feet wide. How much fabric did he use for the curtain?

Reflect

How can you use partial products to multiply 2-digit by 1-digit factors?

Math is... **Mindset**

What has helped you focus when you feel frustrated?

Multiply Multi-Digit by 1-Digit Factors

Be Curious

What do you notice?
What do you wonder?

165 count

165 count

165 count

165 count

165 count

165 count

Math is... **Mindset**

What do you already know that can help you with today's work?

Learn

Solange is playing a video game. She completes 7 levels and scores 125 points at each level.

How many points does Solange score?

You can use partial products to solve the problem.

This equation represents the problem.

$t = 7 \times 125$

Decompose the 3-digit factor by place value.

An area model can help find the partial products.

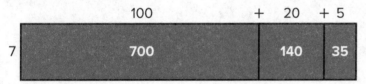

	100	+ 20	+ 5
7	700	140	35

Add the partial products to find the total product.

$t = 700 + 140 + 35$

$t = 875$

Solange scores 875 points in the video game.

> **Math is...** Choosing Tools
>
> How does the area model help you write partial products?

One way to find a product of a multi-digit number and a 1-digit number is to decompose the multi-digit number by place value, find the partial products, and then add the partial products.

⦿ Work Together

Solange starts a new video game. She completes 3 levels and scores 4,732 points at each level. How many points does Solange score? Use partial products to show your work.

On My Own

Name _____

Find the product using an area model and partial products. Show your work.

1. 214 × 2

2. 523 × 4

3. 6,232 × 3

4. 3,317 × 5

5. **STEM Connection** Maya has organized her geode collection in 3 sets. There are 689 geodes in each set. How many geodes does she have?

6. Molly earns $372 each week. How much money does she earn after 9 weeks of work?

7. Each day, 895 people visit the museum. How many people visit the museum in 5 days?

8. Error Analysis Karen is finding 7×632. To find the product, she writes $7 \times 632 = 7 + (600 \times 30 \times 2)$. Do you agree with Karen? Explain why or why not.

9. Draw an area model to show how you can use partial products to find 234×65. Then find the product.

10. Extend Your Thinking Write a word problem that may be modeled by the equation, $w = 3 \times 415$. Then use partial products to solve.

⟳ Reflect

How did area models help you find products of multi-digit and 1-digit factors?

Math is...	Mindset

How did you use what you already know to help you with today's work?

Multiply Two Multiples of 10

Be Curious

Tell me everything you can.

Math is... **Mindset**

How can different ideas and viewpoints help you learn better?

Learn

A case contains 30 bottles of juice.

How many bottles are in 50 cases?

You can use multiplication strategies you know to solve the problem.

> ▶ **One Way** Use basic facts and patterns with place value.
>
> $$30 \times 50 = ?$$
>
> You know $3 \times 5 = 15$
>
> $3 \times 50 = 150$
>
> $30 \times 50 = 1,500$
>
> ---
>
> ▶ **Another Way** Use the Associative Property of Multiplication.
>
> Decompose each factor into a multiple of 10.
>
> $$30 \times 50 = ?$$
> $$3 \times 10 \times 5 \times 10 = ?$$
>
> You can rearrange the factors.
>
> $$3 \times 5 \times 10 \times 10 = ?$$
> $$15 \times 100 = 1,500$$
>
> There are 1,500 juice bottles.

> **Math is...** Patterns
>
> What patterns can you use when multiplying with multiples of 10?

You can use patterns with place value and properties of multiplication to multiply two multiples of 10.

Work Together

> How can you solve the equation? Show your work.
>
> $$60 \times 40 = ?$$

On My Own

Name _____

How can you find the product? Complete the equation.

1. $20 \times 20 = 20 \times$ _____ tens

$=$ _____ tens

$=$ _____

2. $50 \times 40 = 50 \times$ _____ tens

$=$ _____ tens

$=$ _____

3. $70 \times 20 = 7 \times 10 \times$ _____ $\times 10$

$= 7 \times$ _____ $\times 10 \times 10$

$=$ _____ $\times 100$

$=$ _____

4. $90 \times 50 = 9 \times 10 \times$ _____ $\times 10$

$=$ _____ $\times 5 \times 10 \times 10$

$=$ _____ $\times 100$

$=$ _____

5. $90 \times 90 =$ _____

6. $70 \times 50 =$ _____

7. $20 \times 90 =$ _____

8. $20 \times 60 =$ _____

9. A package of pencils contains 20 pencils. How many pencils are in 50 packages?

10. Tisha has 90 dimes. How much money does she have in dollars?

11. Samson exercised for 40 minutes each day for 30 days. How many total minutes did he exercise? Show and explain two ways to solve the problem.

12. **Error Analysis** Alex is using 6 × 5 to find 60 × 50. He says there should be two zeros in the product. Do you agree? Explain why or why not.

13. **Extend Your Thinking** How can you find 30 × 800? Show your work.

Reflect

How did you use patterns while multiplying two multiples of 10?

Math is... **Mindset**

How have different ideas and viewpoints helped you learn better?

Multiply Two 2-Digit Factors

Be Curious

How are they the same?
How are they different?

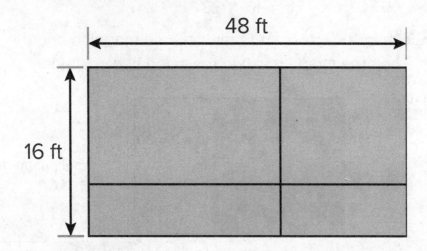

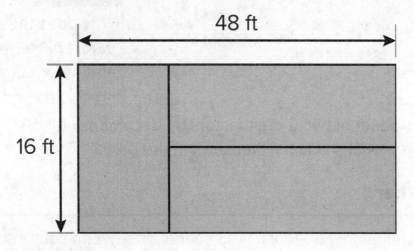

Math is... ▸ **Mindset**

How can working as a
team help you achieve
your goal?

Learn

Paul's classroom is getting new carpet.

How many square feet of carpet does the room need?

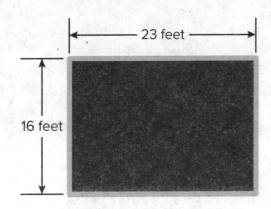

An area model and partial products can help you solve the problem.

This equation represents the problem.

$$c = 16 \times 23$$

You can decompose both factors and draw an area model to find the partial products. The area model has two rows and two columns.

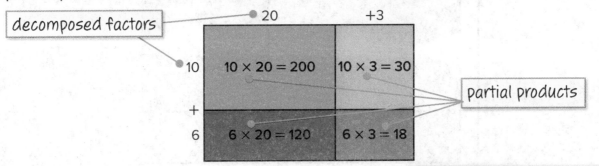

Add the partial products.

$$200 + 30 + 120 + 18 = 368$$

The room needs 368 square feet of carpet.

Math is... Thinking

Why does the area model have two rows and two columns?

You can find the product of two 2-digit factors by decomposing both factors and finding partial products, then adding the partial products.

Work Together

Bay Meadows Elementary School has 15 school buses. Each bus can carry 66 students. How many students can ride a bus to school? Use an area model to show using partial products to solve.

On My Own

Name _____

How can you use partial products to solve? Show your work.

1. $98 \times 20 = ?$

2. $42 \times 38 = ?$

3. _____ $\times$ _____ $= (70 \times 50) + (70 \times 7) + (4 \times 50) + (4 \times 7)$

4. Alvin's Aquarium Shop has 27 fish tanks. Each tank can hold up to 45 fish. What is the greatest number of fish that can be kept at Alvin's Aquarium Shop?

---

5. **STEM Connection** Maya has 13 packs of quartz crystals. Each pack contains 17 crystals. How many crystals does she have?

6. **Error Analysis** To find the product of 27 × 83, Kim writes 27 × 83 = (20 + 80) × (20 + 3) × (7 + 80) × (7 + 3). Do you agree with Kim? Explain why or why not.

7. Tyrone is using (60 × 50) + (60 × 9) + (4 × 50) + (4 × 9) to find the product of two 2-digit factors by using partial products. What two factors could he be multiplying? Explain how you know.

8. **Extend Your Thinking** How can you use an area model to multiply 593 × 42? Use drawings and words to explain your work.

9. There were 45 entrants at the local gymnastics meet. There were 65 times as many entrants at the state gymnastics meet. How many entrants were at the state meet? Use partial products to solve.

Reflect

How did area models help you find products of two 2-digit factors?

Math is... Mindset

How did working as a team help you achieve your goal?

Unit 6
Estimate Products

Name _____

Use reasoning to choose the closest estimate of the product.

1. Circle the best estimate:

23×48

a.	80	b.	800
c.	100	d.	1,000
e.	20	f.	200

Explain your reasoning:

2. Circle the best estimate:

62×29

a.	60	b.	600
c.	120	d.	1,200
e.	180	f.	1,800

Explain your reasoning.

Use reasoning to choose the closest estimate of the product.

3. Circle the best estimate:

   68 × 71

a. 500	b. 5,000
c. 400	d. 4,000
e. 70	f. 140

   Explain your reasoning.

4. Circle the best estimate:

   89 × 68

a. 90	b. 180
c. 400	d. 4,000
e. 600	f. 6,000

   Explain your reasoning.

Reflect On Your Learning

I'm confused. — I'm still learning. — I understand. — I can teach someone else.

Be Curious

What question could you ask?

Math is... Mindset

What helps you understand a problem situation?

Learn

Mr. Gomez purchases 20 cans of paint and 18 brushes to paint a mural. A can of paint costs $11. Brushes cost $3 each.

How much did Mr. Gomez spend on paint and brushes?

Some problems need many steps to get to the solution.

Step 1: Determine the cost of 20 cans of paint.

$20 \times 11 = p$

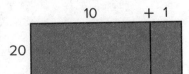

$200 + 20 = 220$

20 cans of paint cost $220.

Step 2: Determine the cost of 18 brushes.

$18 \times 3 = b$

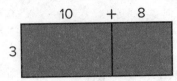

$30 + 24 = 54$

18 brushes cost $54.

> **Math is...** **Explaining**
>
> How can you explain the strategy you use to find partial products?

Step 3: Add to determine the total cost.

$220 + 54 = 274$

20 cans of paint and 18 brushes cost $274.

You can use multiplication strategies you know to help you solve multi-step word problems.

⟲ Work Together

Mrs. Smith's class goes on a field trip to the museum. Tickets to the museum are $6 for children and $18 for adults. What is the total cost for 32 children and 8 adults?

On My Own

Name _____

What are the steps to solve the problem?

1. The cost of an adult movie ticket is $11, and the cost of a children's movie ticket is $8. What is the total cost of 6 adult tickets and 13 children's tickets?

2. Mr. Gabriel purchases supplies for the school store. A notebook costs $3. A package of highlighters costs $4. What is the total cost of 15 notebooks and 12 packages of highlighters?

3. For the morning snack, there are 4 cases of raisins and 6 cases of cheese snacks. Each case of raisins contains 24 boxes, and each case of cheese snacks contains 18 bags. What is the total number of snacks?

4. Barak has $200 to spend on fabric to make curtains. Barak buys 8 yards of velvet. A yard of velvet costs $19. How much money does Barak have left after buying the fabric?

5. Karan makes beaded jewelry bracelets. A bag of beads costs $18. Clasps cost $3 for each bracelet. What is the total cost of 4 bags of beads and 22 clasps?

6. Leona borrows $25 from her mother to buy supplies to sell lemonade. She sells each cup of lemonade for $3. She sells 24 cups of lemonade. How much money did Leona make after paying back her mother for supplies?

7. **Extend Your Thinking** Write a multi-step word problem that could be represented using $(11 \times 2) + (15 \times 5)$. Then solve.

8. **Error Analysis** Sona has 7 boxes of thank-you cards and 9 boxes of get-well cards. There are 14 cards in each box of thank-you cards, and 18 cards in each box of get-well cards. How many cards does Sona have?

Marnie wrote this equation to determine the number of thank-you cards. Do you agree with Marnie? Explain.

$$c = (7 \times 9)$$

Reflect

How can you solve multi-step word problems involving multiplication?

Math is... Mindset

What helped you understand a problem situation today?

Unit Review Name _____

Vocabulary Review

Choose the correct word(s) to complete the sentence.

area model	partial products
decompose	product
Distributive Property	variable
factor	

1. Writing 4 × (10 + 8) as (4 × 10) + (4 × 8) shows application of the
    _____. (Lesson 6-3)

2. In the expression 7 × 9, the number 7 is a(n) _____.
    (Lesson 6-3)

3. A(n) _____ may be used to find partial products.
    (Lesson 6-5)

4. In the equation c = 32 × 16, c is a(n) _____. (Lesson 6-8)

5. The partitioned areas inside an area model represent
    _____. (Lesson 6-4)

6. To _____ 26 by place value means to write it as
    20 + 6. (Lesson 6-3)

7. Given 32 × 46 = 1,472, the number 1,472 is the _____.
    (Lesson 6-3)

Review

8. $700 \times 8 = ?$ (Lesson 6-1)

_____

9. $4 \times 6,000 = ?$ (Lesson 6-1)

_____

10. Complete the equation. (Lesson 6-6)

$70 \times 40 = 70 \times$ _____ tens

$=$ _____ tens

$=$ _____

11. A shelf contains 142 boxes of erasers. Each box contains 4 erasers. About how many erasers are on the shelf? (Lesson 6-2)

12. Carla plans to save $289 each month for 6 months. About how much money will she save?

(Lesson 6-2)

13. How can you decompose a factor to multiply? Draw an array to represent the multiplication.

8×6 (Lesson 6-3)

14. How can you decompose a factor to multiply? Draw an array to represent the multiplication.

7×7 (Lesson 6-3)

15. Elijah builds a raised garden that is 9 yards wide and 32 yards long. What is the area of the garden? Use an area model and partial products to solve. (Lesson 6-4)

16. Mandy receives 1,375 points for each level completed in an online math game. She completes 6 levels. How many points does she receive? Use partial products to solve. (Lesson 6-5)

17. There are 28 almonds in a serving size. How many almonds are in 14 servings? Use an area model and partial products to solve. (Lesson 6-7)

19. A photographer has 4 USB drives that each cost $8.00. He places 12 folders on each USB drive. Each folder contains 65 photos. How many photographs has the photographer put on the USB drives? Write an equation with a variable to solve. (Lesson 6-8)

18. Mrs. Thompson has 35 boxes of colored pencils. Each box contains 12 colored pencils. She gives 8 colored pencils to Mr. Scott. How many colored pencils does she have left? Use an area model and partial products to solve. (Lesson 6-8)

20. There are 28 students in a fourth grade classroom. Each student takes a quiz with 25 questions. During lunch, the teacher checks answers on 7 of the quizzes. How many questions does the teacher have left to check? Explain your answer. (Lesson 6-8)

Performance Task

Maya plans to showcase her gemstone collection in the school auditorium.

Part A: Maya plans to set up 28 tables in the auditorium. Each table will take up 55 square feet of space. How many square feet of space will Maya need?

Part B: The school will allow Maya to use 1,500 square feet of space. Is there enough space for all of Maya's tables? Explain.

Part C: What recommendation would you make to Maya, regarding the number of tables included in her gemstone presentation? Explain.

Reflect

What strategies can be used to multiply multi-digit numbers?

Fluency Practice

Name _____

Fluency Strategy

You can break apart factors to multiply by 7 and 9.

Break apart 7.
$7 = 5 + 2$
$6 \times 7 = 6 \times 5 + 6 \times 2$
$6 \times 7 = 30 + 12$
$6 \times 7 = 42$

Break apart 9.
$9 = 5 + 4$
$4 \times 9 = 4 \times 5 + 4 \times 4$
$4 \times 9 = 20 + 16$
$4 \times 9 = 36$

1. Find 4×7.

Since $7 = 5 +$ _____,

$4 \times 7 = 4 \times$ _____ $+$ _____ $\times 2$

$4 \times 7 =$ _____ $+$ _____

$4 \times 7 =$

Fluency Flash

Use the models to complete the multiplication facts.

2.

$3 \times 7 = 3 \times$ _____ $+$ _____ $\times 2$

$3 \times 7 =$ _____ $+$ _____

$3 \times 7 =$ _____

3.

$2 \times 9 = 2 \times$ _____ $+$ _____ $\times 4$

$2 \times 9 =$ _____ $+$ _____

$2 \times 9 =$ _____

Fluency Check

Find the product.

4. $7 \times 7 =$ _____

5. $3 \times 11 =$ _____

6. $9 \times 6 =$ _____

7. $10 \times 4 =$ _____

8. $7 \times 4 =$ _____

9. $3 \times 6 =$ _____

10. $5 \times 8 =$ _____

11. $4 \times 8 =$ _____

12. $5 \times 7 =$ _____

13. $6 \times 4 =$ _____

14. $9 \times 9 =$ _____

15. $8 \times 8 =$ _____

16. $8 \times 9 =$ _____

17. $7 \times 8 =$ _____

Fluency Talk

How is breaking apart 7 like breaking apart 9 to find a product?

How can you use a 3s fact to find a 6s fact?

Division Strategies with Multi-Digit Dividends and 1-Digit Divisors

Focus Question

How can I divide with multi-digit numbers?

Hi, I'm Finn.

I want to be a construction manager. It is important that work crews each get the same amount of material. It is important for a construction manager to be able to divide!

IGNITE!

Name _____

Equal Shares?

1. Look at your base-ten block representation of 223. Estimate the value of the blocks each member of your group would receive if you shared the collection equally. Record your estimate.

2. Now, distribute, or share, the base-ten blocks equally to each group member. Record the value of each person's blocks.

3. Return all the blocks into one collection. Share the blocks again, but this time leaving one group member out. Record the value of each person's blocks.

4. Return all the blocks into one collection and add a thousands cube. Estimate the value of the blocks each member of your group would receive if you shared the collection equally. Then, share your collection equally and record the results.

Be Curious

What do you notice?
What do you wonder?

Math is... Mindset

What do you do to stay focused on your work?

Learn

How can the solution to 24 ÷ 3 = ? help you determine the solution to 24,000 ÷ 3 = ?

You can use different strategies to solve these equations.

▶ **One Way** Use basic facts and place value.

dividend **divisor** **quotient**

$$24 \div 3 = 8$$

24 tens ÷ 3 = 8 tens   so,   240 ÷ 3 = 80

24 hundreds ÷ 3 = 8 hundreds   so,   2,400 ÷ 3 = 800

24 thousands ÷ 3 = 8 thousands   so, 24,000 ÷ 3 = 8,000

Math is... **Patterns**

How can you use the pattern with zeros to divide other multiples of 10, 100, or 1,000?

▶ **Another Way** Use the relationship between multiplication and division.

3 × 8 = 24	so, 24 ÷ 3 = 8
3 × 80 = 240	so, 240 ÷ 3 = 80
3 × 800 = 2,400	so, 2,400 ÷ 3 = 800
3 × 8,000 = 24,000	so, 24,000 ÷ 3 = 8,000

Patterns exist in the quotients of multiples of 10, 100, and 1,000 that are related to basic facts. The number of zeros in the quotient increases as the number of zeros in the dividend increases.

⬮ Work Together

Miguel bought 300 seedlings for his garden. He wants to plant an equal number of seedlings in 6 rows. How many seedlings go in each row? Show and explain your work.

On My Own

Name _____

How can you complete the equations?

1. 36 ones ÷ 9 = _____ ones

36 tens ÷ 9 = _____ tens

36 _____ ÷ 9 = 4 hundreds

2. 180 ÷ 3 = _____

1,800 ÷ 3 = _____

18,000 ÷ 3 = _____

What is the quotient? Use a related multiplication equation to solve.

3. 48 tens ÷ 6 = ?

6 × 8 _____ = 48 tens

So, 48 tens ÷ 6 = _____

4. 35,000 ÷ 5 = ?

5 × _____ = 35,000

So, 35,000 ÷ 5 = _____

5. 560 ÷ 7 = _____

6. 360 ÷ 4 = _____

What division equation can you use to solve the problem?

7. A bus travels 3,000 miles in 5 days. It travels the same distance each day. How far does the bus travel in one day?

8. A school orders 420 math textbooks. The textbooks arrive in 6 boxes with an equal number of books in each box. How many books are in each box?

9. Naomi reads the same number of pages each day. After 8 days she has read 320 pages. How many pages does she read each day?

10. **STEM Connection** A construction team uses 7,200 nails to build 8 storage buildings. They use the same number of nails for each building. How many nails do they use for each building?

11. Jack watched a movie over 5 equal time periods. The movie was 150 minutes long. For how many minutes did he watch during each time period?

12. Sonia logged her swim times for 3 months. She swam the same amount of time each month and logged 210 hours. How many hours did she swim each month?

13. **Extend Your Thinking** How can you explain the process used to find the quotient of 2,400 ÷ 60?

Reflect

How can you use place value to help you divide multiples of 10, 100, and 1,000?

Math is... **Mindset**
What did you do to stay focused on your work?

Be Curious

How are they the same?
How are they different?

$$25 \div 5 = 5$$
$$250 \div 5 = 50$$
$$2{,}500 \div 5 = 500$$

Math is... Mindset

What behaviors show
respect towards others?

Learn

A hardware store owner is putting nuts and bolts into separate boxes. She divides the nuts equally into 6 boxes and the bolts equally into 7 boxes.

About how many bolts will go in each box?

You can write division equations and estimate using a **range** to solve.

561 nuts   3,358 bolts

In order to estimate the quotient you can use compatible numbers.

Use a compatible number less than the dividend.	Use a compatible greater than the dividend.
$3,358 \div 7 = ?$ ↓ ↓ $2,800 \div 7 = 400$	$3,358 \div 7 = ?$ ↓ ↓ $3,500 \div 7 = 500$

A range of 400 to 500 bolts will go in each box. Since 3,500 is closer to 3,358 than 2,800, the actual quotient will be closer to 500 than 400.

Compatible numbers and related basic facts can be used to find estimated quotients, or a range for the estimated quotients.

> **Math is... Thinking**
>
> How do you know if an estimate will be greater or less than the exact quotient?

ⓒ Work Together

About how many nuts will the hardware owner put in each box?

On My Own

Name _____

How can you estimate the quotient using compatible numbers?

1. $342 \div 8$

2. $836 \div 9$

3. $2{,}134 \div 7$

4. $5{,}361 \div 6$

How can you estimate a range for the quotient?
Write equations to show your work.

5. $749 \div 8$

6. $522 \div 7$

7. $3{,}297 \div 8$

8. $6{,}428 \div 9$

9. A class collected 323 cans for recycling. They can place only 7 cans in each bag. About how many bags will the class need for their cans?

10. Jeremy scored a total of 6,128 points playing video games. If he scored about the same number of points in each of his 9 games, about how many points did he score in each game?

11. The bowling alley had 397 bowlers over the weekend. There were about 5 bowlers for each lane rental. About how many lane rentals did they have for the weekend?

12. **STEM Connection** A construction company has 776 sheets of plywood. They distribute the same number of sheets to 8 construction sites. About how many sheets of plywood will each site receive?

13. How can 24 ÷ 3 help you estimate 238 ÷ 3? Justify your reasoning.

14. **Error Analysis** Louise is using compatible numbers to estimate the quotient of 4,219 ÷ 3. She says that, since 4,219 is between 3,000 and 6,000, the quotient is between 1,000 and 2,000. Do you agree with Louise? Are there other compatible numbers that would work better to estimate the quotient?

15. **Extend Your Thinking** Explain how the process for finding an estimated quotient of 34,219 ÷ 22 compares to the process used to find the estimated quotient of other problems in this lesson. Then find an estimated quotient.

🕐 Reflect

How did you determine the compatible numbers to use when estimating quotients?

Be Curious

Tell me everything you can.

Five students started a sticker club. So far, they have collected some stickers and want to share them equally.

Math is... **Mindset**

How do you feel about learning math?

Learn

Ralph has 42 avocados. He places an equal number into each of three crates.

How many avocados will he put into each crate?

A division equation can represent the problem.

You can use counters and groups to represent the problem.

$$42 \div 3 = c$$

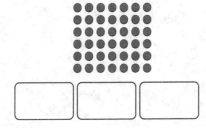

> **Math is... Connections**
>
> What other operation represents equal sharing?

Partition the counters into three equal groups.

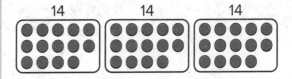

There are 14 counters in each group.

$$42 \div 3 = \mathbf{14}$$

Ralph will put 14 avocados into each crate.

You can use the equal sharing meaning of division to divide a 2-digit number by a 1-digit number.

🗨 Work Together

Shannon uses 52 beads to make 4 bracelets. Each bracelet has the same number of beads. How many beads are in each bracelet?

On My Own

Name _____

How can you solve the problem? Use counters or draw a picture to show your work.

1. 12 counters are shared equally into 3 groups.

There are _____ counters in each group.

2. 25 counters are shared equally into 5 groups.

There are _____ counters in each group.

3. 49 ÷ 7 = _____

4. 39 ÷ 3 = _____

5. 66 ÷ 6 = _____

6. 75 ÷ 5 = _____

7. There are 91 students in the school chorus. The chorus conductor puts 7 students in each row. How many rows of students are there?

8. Four students equally share 68 binder clips. How many binder clips does each student receive?

9. Sasha scores 96 points in 6 games of basketball. She scores the same number of points in each game. How many points does she score in each game?

10. Raul uses 72 nails to build 3 drawers. He uses the same number of nails for each drawer. How many nails does he use for each drawer?

11. **Error Analysis** Marcie says $84 \div 4 = 20$. Do you agree or disagree? Explain your reasoning.

12. **Extend Your Thinking** Mr. Smith has 92 apples to distribute equally into 4 bins. If he distributes 1 apple to each of the 4 bins 10 times, how many apples will be left to distribute?

Reflect

How can you explain the process of dividing by using equal shares?

Math is... Mindset

How did learning math make you feel?

Understand Partial Quotients

Be Curious

Tell me everything you can.

Math is... Mindset

How can you make sure you share your thinking clearly?

Learn

The local market has 105 bags of pecans.
They put an equal number of bags on 5 tables.
How many bags of pecans are on each table?

You can use division to determine how many bags are on each table.

Base-ten blocks can represent the problem.

$105 \div 5 = ?$

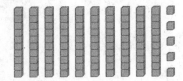

You can use the **partial quotients** strategy to divide. You use compatible numbers to determine the number of 5s in 105.

20 is a partial quotient.

$$
\begin{array}{r}
105 \\
-100 \quad (5 \times 20) \\
\hline
5 \\
- \quad 5 \quad (5 \times 1) \\
\hline
0
\end{array}
$$

$20 + 1 = 21$
$105 \div 5 = 21$
Each table has 21 bags of pecans.

Partial quotients is one strategy to divide a 3-digit number by a 1-digit number.

Math is... Connections

How is multiplication related to the partial-quotients strategy?

Work Together

How can you use partial quotients to solve the problem?

$216 \div 9 = ?$

On My Own

Name _____

What is the quotient? Use a representation to show the partial quotients.

1. $136 \div 8 =$ _____

2. $114 \div 6 =$ _____

3. $115 \div 5 =$ _____

4. $105 \div 3 =$ _____

What is the quotient? Use the partial-quotients strategy to solve.

5. $154 \div 7 =$ _____

6. $342 \div 9 =$ _____

7. Will stacked 135 quarters. He put 9 quarters into each stack. How many stacks did he make?

8. Jeremy put 256 baseball cards into 8 binders. Each binder had the same number of baseball cards. How many baseball cards were in each binder?

9. There are 210 workers at the football stadium to help clean up after the game. The workers are divided into 5 equal teams. How many workers are on each team?

10. Deborah is making bead necklaces for her friends. She uses 306 beads for 9 necklaces. She uses the same number of beads for each necklace. How many beads does Deborah use for each necklace?

11. **Error Analysis** Marsha says she can use 10 as the first three partial quotients when finding 261 ÷ 9. Do you agree or disagree? Explain your reasoning.

12. **Extend Your Thinking** How can you find 316 ÷ 4 two different ways by using different partial quotients in each solution?

Reflect

How can you decide what number to use as a partial quotient?

Math is... Mindset

What did you do to share your thinking clearly?

Divide 4-Digit Dividends by 1-Digit Divisors

Be Curious

How are they the same?
How are they different?

	10	10	10	10	2
4	40	40	40	40	8

	40	2
4	160	8

Math is... **Mindset**

How can you show that you understand your partner's point of view?

Learn

A factory is packaging 1,550 golf balls.
Each box holds 5 golf balls.

How many boxes will the factory fill?

You can use an area model to represent and solve the problem.

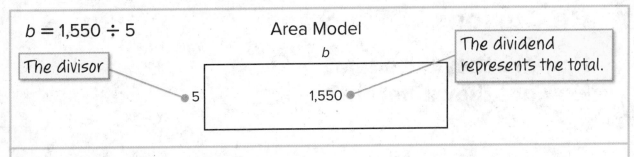

$b = 1,550 \div 5$

The divisor

Area Model

b

5   1,550

The dividend represents the total.

Think: How many 5s are in 1,550?

Area Model

100 + 100 + 100 + 10

5 | 500 | 500 | 500 | 50

Partial Quotients

$$
\begin{array}{rl}
1,550 & \\
-\ \ 500 & (5 \times 100) \\
\hline
1,050 & \\
-\ \ 500 & (5 \times 100) \\
\hline
550 & \\
-\ \ 500 & (5 \times 100) \\
\hline
50 & \\
-\ \ 50 & (5 \times 10) \\
\hline
0 &
\end{array}
$$

Add the partial quotients to find the final quotient.

$100 + 100 + 100 + 10 = 310$

$b = 1,550 \div 5$

$b = 310$

The factory can fill 310 boxes of golf balls.

Math is... **Patterns**

What other partial quotients could you use?

You can use partial quotients to divide 4-digit dividends by 1-digit divisors.

Work Together

How can you use partial quotients to solve the equation?

$4,564 \div 4 = ?$

On My Own

Name _____

What is the quotient? Use the partial quotients to solve.

1. $2,200 \div 2 =$ _____

$$
\begin{array}{r}
2,2\,0\,0 \\
-\ 2,0\,0\,0\ (2 \times 1,000) \\
\hline
2\,0\,0 \\
-\ 2\,0\,0\ (2 \times 100) \\
\hline
0
\end{array}
$$

2. $4,840 \div 4 =$ _____

$$
\begin{array}{r}
4,8\,4\,0 \\
-\ 4,0\,0\,0\ (4 \times 1,000) \\
\hline
8\,4\,0 \\
-\ 4\,0\,0\ (4 \times 100) \\
\hline
4\,4\,0 \\
-\ 4\,0\,0\ (4 \times 100) \\
\hline
4\,0 \\
-\ 4\,0\ (4 \times 10) \\
\hline
0
\end{array}
$$

What is the quotient? Use partial quotients to solve.

3. $9,300 \div 3 =$ _____

4. $3,240 \div 3 =$ _____

5. $3,216 \div 2 =$ _____

6. $8,350 \div 5 =$ _____

7. There are 1,359 students attending field day. There are 9 different game stations. Each game station holds the same number of students. How many students will be at each game station?

8. Zoe bought 2,268 inches of ribbon. She is making 4 different costumes with each costume using the same amount of ribbon. How many inches of ribbon will Zoe use for each costume?

9. A dairy farm processes 5,112 gallons of milk in a 9-day period. They process the same amount of milk each day. How many gallons of milk do they process each day?

10. The state fair sold 6,834 tickets in 3 weeks. The same number of tickets were sold each week. How many tickets were sold each week?

11. **Error Analysis** Ralph says he can use 1,000 as the first partial quotient to solve 1,962 ÷ 9. How can you respond to Ralph?

12. **Extend Your Thinking** You can use 100 as the first partial quotient to find 2,316 ÷ 4. What are two other first partial quotients you could use?

Reflect

How do you use the partial quotients process to find quotients of 4-digit dividends and 1-digit divisors?

Math is... Mindset

How did you show you understand your partner's point of view?

Be Curious

What could the question be?

Math is... **Mindset**

How can you contribute to a productive classroom culture?

Learn

Avery's classroom library has 235 books in one bookcase. The bookcase has 8 shelves. Each shelf has the same number of books on it.

How many books are on each shelf?

You can divide to solve the problem.

$235 \div 8 = ?$

Use the partial quotients strategy.

$$\begin{array}{r} 235 \\ -160 \quad (8 \times 20) \\ \hline 75 \\ -72 \quad (8 \times 9) \\ \hline 3 \end{array}$$

> Think: What number times 8 is closest to 235?

> a remainder

235 cannot be divided evenly into 8. There is a **remainder** of 3.

Add the partial quotients and write the remainder.

$235 \div 8 = 29 \text{ R}3$

> The remainder can be shown with the letter R.

There are 29 books on each shelf with 3 books left over.

Math is... Connections

In the partial-quotients strategy, what indicates that there is a remainder?

When whole numbers cannot be divided evenly, the quotient will include a remainder.

Work Together

Avery's school purchased 7,220 books for the students. If each student receives 6 books, how many books will be left over? Use the partial-quotients strategy to solve the problem.

On My Own

Name _____

What is the quotient? Use the partial quotients to solve.

1. 415 ÷ 2 = _____ R _____

$$
\begin{array}{r}
4\,1\,5 \\
-\,2\,0\,0 \\ \hline
2\,1\,5 \\
-\,2\,0\,0 \\ \hline
1\,5 \\
-\,1\,4 \\ \hline
1
\end{array}
\quad
\begin{array}{l}
 \\
(2 \times 100) \\
 \\
(2 \times 100) \\
 \\
(2 \times 7) \\
\end{array}
$$

2. 5,044 ÷ 5 = _____ R _____

$$
\begin{array}{r}
5{,}0\,4\,4 \\
-\,5{,}0\,0\,0 \\ \hline
4\,4 \\
-\,4\,0 \\ \hline
4
\end{array}
\quad
\begin{array}{l}
 \\
(5 \times 1{,}000) \\
 \\
(5 \times 8) \\
\end{array}
$$

What is the quotient and the remainder? Use partial quotients to solve.

3. 929 ÷ 3 = _____

4. 119 ÷ 4 = _____

5. 3,225 ÷ 8 = _____

6. 8,254 ÷ 5 = _____

7. 8,437 ÷ 7 = _____

Solve the problem.

8. A restaurant has $609 to buy cups. If each box of cups costs $9, how many boxes can the restaurant purchase? How much money will be left over?

9. A party planner has 275 balloons for a party. How many tables can he have with 6 balloons on each table? How many balloons will be left over?

10. George has $20 and wants to buy snow cones for his friends. The snow cones are $3 each. How many snow cones can he buy? How much money will he have left?

11. When you solve a division problem, how do you know if you have a remainder?

12. **Extend Your Thinking** For the dividend 1,240, how do you know that dividing by 2, 5, and 10 will not result in a remainder?

Reflect

How can you make sense of a remainder while solving a word problem involving division?

Math is... **Mindset**

How did you contribute to a productive classroom culture today?

Make Sense of a Remainder

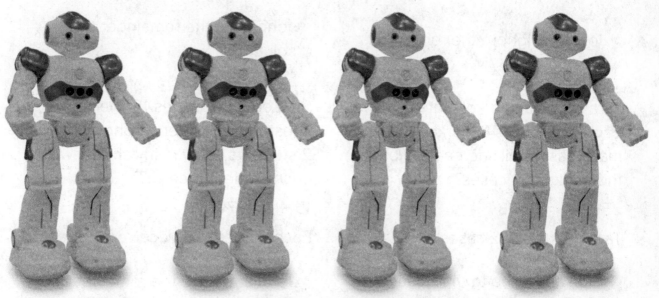

Be Curious

What could the question be?

4 batteries required   4 batteries required   4 batteries required   4 batteries required

14 Pack

POWER BATTERY

Math is... **Mindset**

What helps you focus when you feel frustrated?

Learn

John is solving the following division problems.

How can he interpret the remainder in each problem?

Jia-Li puts 98 scissors into boxes. A box holds 4 scissors. How many boxes can she fill?

$98 \div 4 = 24 \text{ R}2$

Jia-Li can fill 24 boxes.

> John can ignore the remainder.

Raj puts 38 paint jars into boxes. A box holds 8 jars. How many jars will be left over?

$38 \div 8 = 4 \text{ R}6$

There will be 6 jars left over.

> John uses the remainder as the solution.

Tara puts 251 markers into cases. Each case holds 8 markers. How many cases will she need to put all the markers in cases?

$251 \div 8 = 31 \text{ R}3$

Tara will need 32 cases.

> John adds one to the quotient for the solution.

Sam equally divides 15 sheets of construction paper among 2 students. How many sheets will each student receive?

$15 \div 2 = 7 \text{ R}1$

Each student will receive $7\frac{1}{2}$ sheets.

> John can write the remainder as a fraction.

When a division problem has a remainder, you interpret the remainder to determine the solution to the problem.

Math is... Explaining

How can you use the situation to explain what the remainder represents?

Work Together

Forty-one people want to take a tour of the city. The tour guides can only take 6 people in a group. How many tour guides are needed? Explain your solution.

How can you solve the problem? Show your work.

1. Caleb makes fruit smoothies. He has 26 strawberries. If Caleb puts 4 strawberries into each smoothie, how many smoothies can he make? How many strawberries will be left over?

2. There are 48 ounces of water in a pitcher. How many 10-ounce bottles can Sven fill using the pitcher?

3. There are 125 chairs to put in rows. Each row can have 20 chairs. How many rows are needed for all the chairs?

4. Herbert has 147 postcards. He places 6 postcards on a page in his album. How many pages will he need for all his postcards?

5. Two families will share 5 oranges at a picnic. How many oranges will each family receive if they share all the oranges equally?

For exercises 6 and 7, choose the answer that will correctly describe the effect of the remainder for each situation.

6. **STEM Connection** A truck can hold 8 crates of building materials. How many trips will the truck need to make to get 65 crates to the construction site?

   A. The remainder is the solution.

   B. The remainder can be ignored.

   C. The remainder forces the quotient to be the next whole number.

   D. The remainder is partitioned as a fraction.

7. A ribbon is 25 inches long. How many 6-inch pieces can be made?

A. The remainder is the solution.

B. The remainder can be ignored.

C. The remainder forces the quotient to be the next whole number.

D. The remainder is partitioned as a fraction.

8. A van can hold 9 people. How many trips are needed to get 33 people to the airport?

9. There are 94 baseball cards to share equally among 4 friends. How many baseball cards will each friend get?

10. Error Analysis Roger says a remainder always forces the answer to be the next whole number. How can you explain to Roger that his statement is not always true?

11. Extend Your Thinking For the equation $17 \div 8 = ?$, create a situation in which the remainder can be ignored.

🕐 Reflect

How can a remainder affect the answer to a problem?

Math is... Mindset

How have you stayed focused when you were frustrated?

Interpreting Remainders

Name _____

Read each problem. Choose the answer that best represents the situation.

1. Diana has 29 photos to put in a photo album. She can fit 6 photos on a page. What is the fewest number of pages she needs in order to fit all the photos in the album?

Which number of pages best represents the situation? Circle your choice.	Explain your choice.
A. 4 **B.** 5 **C.** 4 with 5 left over **D.** 6	

2. Anna has 62 grapes. She wants to share them equally among 7 friends. Anna will only eat any extras that are left over. How many grapes will Anna eat?

Which number of grapes best represents the situation? Circle your choice.	Explain your choice.
A. 6 **B.** 8 **C.** 8 with 6 left over **D.** 9	

Read the problem. Choose the answer that best represents the situation.

3. Mr. Garcia has 62 markers to share among 8 groups of students. He wants to give out as many markers as possible so that each group gets the same number. How many markers should Mr. Garcia give each group?

Which number of markers best represents the situation? Circle your choice.	Explain your choice.
A. 6	
B. 7	
C. 8	
D. 7 with 6 left over	

Reflect On Your Learning

I'm confused.   I'm still learning.   I understand.   I can teach someone else.

?

Be Curious

What math do you see?

Kim is making bouquets. She will put some roses in each bouquet. She had some roses. She gave some roses to her mother.

Math is... Mindset

How can a plan help you solve a problem?

Learn

Kim is making bouquets. She will put 6 roses in each bouquet.
She had 124 roses. She gave 18 roses to her mother.

How many bouquets can Kim make now?

Some problems have more than one problem to solve.

How many roses does Kim have after giving some to her mother?	How many bouquets can Kim make?
A subtraction equation represents the problem.	A division equation represents the problem.

$r = 124 - 18$

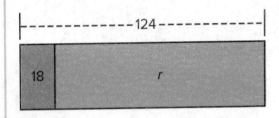

$b = 106 \div 6$

$$\begin{array}{r} 106 \\ -\ 60 \\ \hline 46 \\ -\ 42 \\ \hline 4 \end{array}$$ (6 × 10)

(6 × 7)

$b = 17$ R4

$r = 106$
Kim has 106 roses for the bouquets.

Kim can make 17 bouquets with 6 roses each, with 4 roses left.

Math is... **Perseverance**

What are some other strategies you might try?

For problems with more than one question to answer, you determine which operation can answer each question.

💬 Work Together

Roger is packing bottles of juice into cartons. He can put 9 bottles in each carton. He had 64 bottles of juice, but 12 were damaged. How many cartons will he need to pack all the bottles that are not damaged? Use numbers, words, and representations to show your work.

On My Own

Name _____

Solve each problem.

1. A school storeroom has 235 sticky note packages in one box and 224 in another box. The sticky notes will be equally distributed to 9 classes. How many sticky note packages will each class receive?

2. A florist has 875 flowers. She uses 49 flowers in a bouquet. The rest will be put in vases. Each vase can hold 8 flowers. How many vases will the florist need for the rest of the flowers?

3. Carly has 168 stickers in one book and 492 stickers in another book. There are 6 stickers on each page of the books. How many pages has she filled in the books?

4. Write your own multi-step word problem that can be represented by the equations $2 + 4 = g$ and $204 \div g = ?$ Then solve.

5. **STEM Connection** Finn plans to use boards to build a 78-foot border around his garden. Each board is 3 feet long. The boards are sold in packages of 3. How many packages of boards should Finn order?

6. There are 129 girls and 204 boys going on a hiking trip. There is a maximum of 6 students with each hiking guide. All the students need to be with a guide. How many guides are needed?

7. Karen can make 7 origami frogs each day and has already made 304 frogs. How many days are left until she makes 1,000 frogs?

8. **Extend Your Thinking** The cafeteria needs enough tables to hold 174 students and 13 teachers. The cafeteria tables seat 8 people. Ryan finds how many tables they need. His work is shown. Explain what each number in his work stands for. What is the fewest number of tables they need?

$$(174 + 13) = 187$$
$$187 \div 8 = 23 \text{ R } 3$$

Reflect

Why might you have to interpret a remainder when solving a problem?

Math is... Mindset

How did you use a plan to solve a problem?

Unit Review   Name _____

Vocabulary Review

Choose the correct word(s) to complete the sentence.

area model	compatible number	dividend
divisor	partial quotients	quotient
remainder		

1. In the equation $42 \div 6 = 7$, the number 7 is the _____.
    (Lesson 7-1)

2. A(n) _____ is a number that makes it
    easy to find an estimate. (Lesson 7-2)

3. A(n) _____ is a number that divides another
    number. (Lesson 7-1)

4. A(n) _____ may be used to solve division
    problems. (Lesson 7-5)

5. A quotient may be calculated by finding _____,
    which are partial answers at each step. (Lesson 7-4)

6. A(n) _____ is the number being divided by
    another number. (Lesson 7-1)

7. A(n) _____ is the amount left over after
    dividing. (Lesson 7-6)

Review

8. Andy, Carlos, and Ashley mow lawns to earn extra money. This week, they earned a total of $180. If they divide their earnings equally, how much will each of them receive? (Lesson 7-1)

9. Michelle has 400 blank cards available to use for 8 special events this year. She plans to use an equal number of cards for each special event. How many cards will she use for each event? (Lesson 7-1)

10. What is a reasonable estimate of the quotient?

$446 \div 7 = d$ (Lesson 7-2)

11. How can you use compatible numbers to find an estimated range for the quotient?

$6,845 \div 8$ (Lesson 7-2)

12. A school is holding a bake sale in the gym. There are 378 fruit bars available for sale. If the PTO plans to place about the same number of fruit bars on each of 9 tables, how many fruit bars will be placed on each table? (Lesson 7-3)

13. Megan has 98 bracelets. She puts an equal number of bracelets into 7 boxes. How many bracelets will she put in each box? (Lesson 7-3)

14. There are 78 students in an engineering club. The sponsor puts 6 students in each group. How many groups of students are there? (Lesson 7-3)

15. Josh is solving $28,000 \div 7 = ?$ How many thousands will be in the quotient? What is the quotient? (Lesson 7-1)

16. A bakery has 192 muffins ready to place into boxes. The baker puts an equal number of muffins into 8 boxes. How many muffins are in each box? Use partial quotients to solve. (Lesson 7-4)

```
  192
 −160   (8 × 20)
   32
 − 32   (8 × 4)
    0
```

17. A school has raised $1,650 that will be used to provide funding to 5 teachers. The school plans to give an equal amount of money to each teacher. How much money will each teacher receive? Use partial quotients to solve. (Lesson 7-5)

```
  1,650
 −1,500   (5 × 300)
    150
 −  150   (5 × 30)
      0
```

18. Mandy has 342 pumpkin seeds to put into 8 bags. She puts an equal number of seeds in each bag. Which statement is true? (Lesson 7-6)

A. She can put 42 seeds in each bag and have 6 seeds left over.

B. She can put 43 seeds in each bag and have none left over.

C. She can put 41 seeds in each bag and have 6 seeds left over.

D. She can put 42 seeds in each bag and have 4 seeds left over.

19. Jillian has 37 toy cars to place into some boxes. She can fit 8 toy cars into each box. What is the fewest number of boxes she needs in order to fit all the toy cars into boxes? (Lesson 7-8)

20. Seema is cutting fabric pieces for a quilting project. She cuts 35 yards of fabric equally to make 8 quilts. How much fabric does she use for each quilt? (Lesson 7-8)

Performance Task

Part A: Finn and Sarah are construction managers. They are loading metal beams into the back of a truck. Finn loads 84 beams. Each beam weighs 4 pounds. Sarah's beams weigh 3 pounds each. The total weight of the beams she adds to the back of the truck is the same weight that Finn adds to the truck. How many beams did they load into the back of the truck? Explain your answer.

Part B: They dropped off all the beams they loaded into the back of the truck at construction sites. They dropped off the same number of beams at each construction site and there are less than 10 construction sites. How many construction sites could there be? Explain your answer.

 **Reflect**

What strategies can be used to divide with multi-digit numbers?

Unit 7
Fluency Practice

Name _____

Fluency Strategy

> You can use an **algorithm** to add two whole numbers. Add the numbers in the same place value.
>
> $1{,}524 + 2{,}345 = ?$
>
> Add from right to left. Add the ones, tens, hundreds, then thousands.
>
> $$\begin{array}{r} 1{,}524 \\ +\,2{,}345 \\ \hline 3{,}869 \end{array}$$

1. How can you use an **algorithm** to add $257 + 541 = ?$

Fluency Flash

Write the sum in the place-value chart.

2.

thousands	hundreds	tens	ones
2	3	3	4
+ 1	1	2	3

3.

thousands	hundreds	tens	ones
2	2	7	4
+ 3	2	1	5

Fluency Check

Find the sum or product.

4. $2{,}546 + 1{,}423 =$ _____

5. $7 \times 9 =$ _____

6. $3{,}416 + 1{,}381 =$ _____

7. $3 \times 5 =$ _____

8. $4{,}216 + 1{,}371 =$ _____

9. $5{,}136 + 4{,}712 =$ _____

10. $8 \times 3 =$ _____

11. $6 \times 9 =$ _____

12. $2{,}121 + 7{,}346 =$ _____

13. $9 \times 4 =$ _____

14. $536 + 419 =$ _____

15. $8 \times 7 =$ _____

16. $7 \times 3 =$ _____

17. $8{,}512 + 1{,}346 =$ _____

Fluency Talk

How would you add two numbers that have a different number of digits?

Explain to a friend how you can break apart 9 to multiply.

Glossary/Glosario

English	Spanish/Español

Aa

acute angle An angle with a measure greater than 0° and less than 90°.

ángulo agudo Un ángulo que mide más de 0° y menos de 90°.

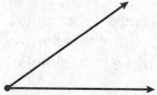

acute triangle A triangle with all three angles less than 90°.

triángulo acutángulo Un triángulo cuyos tres ángulos miden menos de 90°.

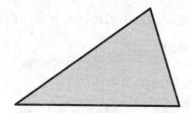

additive comparison Comparing two quantities with addition to determine how much more one is than the other.

comparación aditiva Comparación de dos cantidades de adición a fin de determinar cuánto mayor es una que la otra.

algorithm A way of doing something in math. It is a set of steps that always works if done correctly.

algoritmo Manera de resolver en matemáticas. Es una serie de pasos que si se realizan correctamente, siempre funcionan.

angle A figure that is formed by two rays with the same endpoint.

ángulo Figura formada por dos rayos con el mismo extremo.

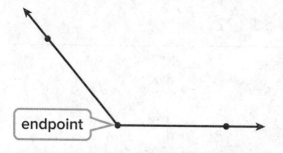

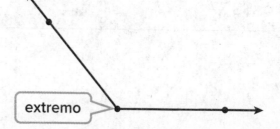

English	Spanish/Español

Associative Property of Multiplication The property which states that the grouping of the factors does not change the product.

$(4 \times 2) \times 3 = 4 \times (2 \times 3)$

propiedad asociativa de la multiplicación Propiedad que establece que la agrupación de los factores no altera el producto.

$(4 \times 2) \times 3 = 4 \times (2 \times 3)$

Bb

benchmark fractions Commonly used fractions that can be used for estimation.

fracciones de referencia Fracciones comúnmente usadas que pueden utilizarse para realizar estimaciones.

Cc

composite number A whole number that has more than two factors.

Example: 12 has the factors 1, 2, 3, 4, 6, and 12.

número compuesto Número entero que tiene más de dos factores.

Ejemplo: 12 tiene a los factores 1, 2, 3, 4, 6 y 12.

cup A customary unit for measuring capacity.

1 cup = 8 ounces
16 cups = 1 gallon

taza Una unidad para medir la capacidad.

1 taza = 8 onzas y 16 tazas = 1 galón

Dd

decimal A number that has one or more digits to the right of the decimal point.

decimal Número con uno o mas digitos a la derecho del punto decimal.

decimal point A period separating the ones and the tenths in a decimal number.

0.8 or $3.77

punto decimal Punto que separa las unidades y las décimas en un número decimal.

0.8 o $3.77

English	Spanish/Español

Distributive Property of Multiplication To multiply a sum by a number, you can multiply each addend by the same number and add the products.

$8 \times (9 + 5) = (8 \times 9) + (8 \times 5)$

propiedad distributiva de la multiplicación Para multiplicar una suma por un número, puedes multiplicar cada sumando por el mismo número y sumar los productos.

$8 \times (9 + 5) = (8 \times 9) + (8 \times 5)$

equiangular triangle A triangle with 3 congruent angles.

triángulo equiangular Un triángulo con 3 ángulos congruentes.

equilateral triangle A triangle with three congruent sides.

triángulo equilátero Triángulo con tres lados congruentes.

equivalent fractions Fractions that have the same value.

fracciones equivalentes Fracciones que tienen el mismo valor.

factor pair The two factors that are multiplied to find a product.

pares de factores Los dos factores que se multiplican para hallar un producto.

fluid ounce A customary unit of capacity.

onza líquida Unidad usual de capacidad.

hundredth A place value position. One of one hundred equal parts.

Example: In the number 0.57, 7 is in the hundredths place.

centésima Valor de posición. Una de cien partes iguales.

Ejemplo: En el número 0.57, 7 está en el lugar de las centésimas.

Ii

isosceles triangle A triangle with at least 2 sides of the same length.

triángulo isósceles Triángulo que tiene por lo menos 2 lados del mismo largo.

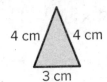

4 cm   4 cm
3 cm

Kk

kilometer A metric unit for measuring length.

1 km = 1,000 m

kilómetro Unidad métrica de longitud.

1 km = 1,000 m

Ll

line of symmetry A line on which a figure can be folded so that its two halves match exactly.

eje de simetría Línea por la que puede doblarse una figura de manera que ambas mitades coincidan exactamente.

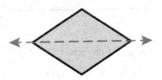

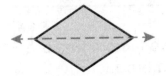

line segment A part of a line between two endpoints. The length of the line segment can be measured.

segmento de recta Parte de una recta entre dos extremos. La longitud de un segmento de recta se puede medir.

A   B

A   B

English	Spanish/Español
line symmetry A line that can be drawn through the figure which splits the figures into 2 halves that match.	**simetría axial** Línea que divide una figura en dos mitades exactamente iguales.

This cube has line symmetry.

Este cubo tiene simetría axial.

Mm

millimeter A metric unit used for measuring length.

1,000 millimeters = 1 meter

milímetro Unidad métrica de longitud.

1,000 milímetros = 1 metro

mixed number A number that has a whole-number part and a fraction part.

$6\frac{3}{4}$

número mixto Número compuesto por una parte entera y una parte fraccionaria.

$6\frac{3}{4}$

multiplicative comparison Comparing two quantities using multiplication where one is a multiple of the other.

comparación multiplicativa Comparación de la multiplicación de dos cantidades donde una es múltiplo de la otra.

Oo

obtuse angles An angle that measures greater than 90° but less than 180°.

ángulo obtuso Ángulo que mide más de 90° pero menos de 180°.

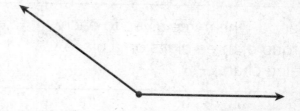

English	Spanish/Español

obtuse triangle A triangle with one obtuse angle.

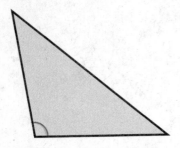

triángulo obtusángulo Triángulo con un ángulo obtuso.

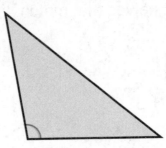

ounce A customary unit for measuring weight or capacity.

onza Unidad inglesa de peso o capacidad.

Pp

parallel lines Lines that are the same distance apart. Parallel lines do not meet.

rectas paralelas Rectas separadas por la misma distancia. Las rectas paralelas no se intersecan.

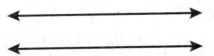

partial products A multiplication method in which the products of each place value are found separately and then added together.

productos parciales Método de multiplicación por el cual los productos de cada valor posicional se hallan por separado y luego se suman entre sí.

partial quotients algorithm A method that shows the partial answer, or quotient, at each step. After all the steps have been completed, all the partial quotients are added to find the final quotient.

algoritmo de cocientes parciales Método que muestra la respuesta parcial, o cociente, en cada paso. Una vez completados todos los pasos, se suman todos los cocientes parciales para hallar el cociente final.

period The name given to each group of three digits on a place-value chart.

período Nombre dado a cada grupo de tres dígitos en una tabla de valores de posición.

perpendicular lines Lines that meet or cross each other to form right angles.

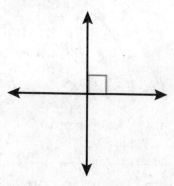

rectas perpendiculares Rectas que se intersecan o cruzan formando ángulos rectos.

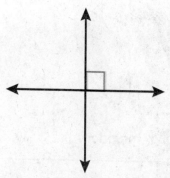

pint A customary unit for measuring capacity.

1 pint = 2 cups

pinta Una unidad de la sistema Inglés para medir la capacidad.

1 pinta = 2 tazas

prime number A whole number with exactly two factors, 1 and itself.

7, 13, and 19

número primo Número entero que tiene exactamente dos factores, 1 y sí mismo.

7, 13, y 19

Qq

quart A customary unit for measuring capacity.

1 quart = 4 cups

cuarto Unidad usual de capacidad.

1 cuarto = 4 tazas

Rr

ray A part of a line that has one endpoint and extends indefinitely in one direction.

semirrecta Parte de una recto que tiene un extremo y que se extiende sin fin en una dirección.

English	Spanish/Español

remainder The number that is left after one whole number is divided by another.

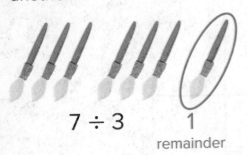

$$7 \div 3 \qquad 1$$
remainder

residuo Número que queda después de dividir un número entero entre otro número entero.

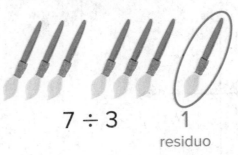

$$7 \div 3 \qquad 1$$
residuo

---

right triangle A triangle with one right angle.

triángulo rectángulo Triángulo con un ángulo recto.

Ss

scalene triangle A triangle with no congruent sides.

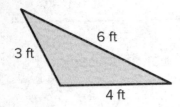
6 ft
3 ft
4 ft

triángulo escaleno Triángulo sin lados congruentes.

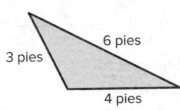
6 pies
3 pies
4 pies

Tt

tenth One of ten equal parts or $\frac{1}{10}$.

décima Una de diez partes iguales ó $\frac{1}{10}$

---

ton A customary unit to measure weight.

1 ton = 2,000 pounds

tonelada Unidad inglesa de peso.

1 tonelada = 2,000 libras